Frank L. (Franklin Leonard) Pope

Modern Practice of the Electric Telegraph

A Technical Handbook for Electricians, Managers, and Operators

Frank L. (Franklin Leonard) Pope

Modern Practice of the Electric Telegraph
A Technical Handbook for Electricians, Managers, and Operators

ISBN/EAN: 9783743687004

Printed in Europe, USA, Canada, Australia, Japan

Cover: Foto ©berggeist007 / pixelio.de

More available books at **www.hansebooks.com**

MODERN PRACTICE

OF THE

ELECTRIC TELEGRAPH

A TECHNICAL HANDBOOK

FOR

ELECTRICIANS, MANAGERS, AND OPERATORS

WITH 185 ILLUSTRATIONS

BY

FRANKLIN LEONARD POPE

PAST PRESIDENT OF THE AMERICAN INSTITUTE OF ELECTRICAL ENGINEERS; MEMBER OF
THE INSTITUTION OF ELECTRICAL ENGINEERS (LONDON)

FIFTEENTH EDITION, REWRITTEN AND ENLARGED

NEW YORK
D. VAN NOSTRAND COMPANY
LONDON
SAMPSON, LOW, MARSTON & CO.
(LIMITED)
1899

COPYRIGHT, 1891
D. VAN NOSTRAND COMPANY

78687

Braunworth, Munn & Barber,
Printers and Binders,
Brooklyn, N. Y.

IN AFFECTIONATE REMEMBRANCE

OF MY FORMER CHIEF

Marshall Lefferts

ENGINEER OF

THE AMERICAN TELEGRAPH COMPANY, 1861–1864

UNDER WHOSE

ENLIGHTENED, PROGRESSIVE, AND LIBERAL ADMINISTRATION THE

METHODS OF MODERN SCIENCE WERE FIRST APPLIED TO

AMERICAN TELEGRAPHY

MORSE (SAMUEL FINLEY BREESE), inventor of the recording electro-magnetic telegraph, born in Charlestown, Mass., April 27, 1791; graduated at Yale, 1810; studied art in the Royal Academy of London, 1811-15, under Benjamin West. In 1829 he again visited Europe for further study of his profession, and while returning home in 1832, on board the ship Sully, conceived and made drawings of his recording telegraph (see J. D. REID: *The Telegraph in America*, chapters vi., vii.). From this time until his death he was unremittingly employed with his invention, passing meantime through many vicissitudes of fortune, and some most painful experiences. He was first professor of fine arts in the University of the City of New York, in one of the rooms of which institution he set up in 1835 his first crude recording telegraphic apparatus, now preserved in the cabinet of the Western Union Company in New York. In 1837, Alfred Vail, a skilful mechanic and inventor, became his partner in the enterprise. Vail entirely reconstructed the apparatus, and embodied it in the practical form in which it was first introduced to the commercial world. After a series of discouragements that would have utterly disheartened most men, Morse, assisted by Vail, established in 1844, under an appropriation from Congress, the first line between Washington and Baltimore. On May 24 of that year, Morse put to the test the great experiment on which his mind had been laboring for many anxious and weary years. His triumph was complete. Honors and riches were showered upon him at home and abroad. Professor Morse was a man of great simplicity of character, firm in his friendships, and most persistent and exhaustive in all his undertakings. He wielded the pen of a ready writer, and his genius, learning, and taste were illustrated by numerous contributions to the press, evincing not only graceful rhetoric but elaborate and well-sustained argument. On June 10, 1871, a bronze statue of Morse, erected by the contributions of the thousands of telegraphic employees in America, was unveiled with imposing ceremonies in Central Park, New York. He died in New York, April 2, 1872.

PREFACE.

ALMOST a quarter of a century has passed since the publication of the first edition of this work. During that period, and more especially during the past ten years, the progress which has been made in the application of electricity to the industrial arts has been literally unprecedented, while the extraordinary practical results which have been attained have exerted a reflex action in stimulating in an equal degree the advancement of electrical science; an advancement which has not been without its influence upon the theory and practice of the electric telegraph. This circumstance has at length rendered necessary, not a mere revision of the original treatise, but the preparation, in fact, of an entirely new work throughout.

To the intelligent and observant mind of youth, the art of telegraphy possesses a singular fascination, and in many instances its pursuit tends to excite a spirit of scientific inquiry, not only commendable in itself, but valuable as establishing a sure foundation for future success in broader fields of labor. It has been the aim of the author to supply a knowledge, not only of the principles and practice of telegraphy, but of the theory of electricity and the methods of electrical measurement, which should be of the highest possible value to every person entrusted with the care and management of telegraphic apparatus. It has, however, been deemed advisable to somewhat restrict the scope of the work, and hence the automatic, type-printing, synchronous, submarine, and other methods, requiring on the part of the practitioners a special training apart from a knowledge of the ordinary system, have been excluded. The construction and maintenance of aerial, subterranean, and submarine lines has also, by a natural process of evolution in the progress of the art, become a separate profession, and the subject can therefore receive but brief notice in a work primarily designed for the guidance and instruction of the practical operator.

In the treatment of the subject, the use of mathematics has been rendered unnecessary by the free introduction of concrete examples, illustrative of methods and processes of arithmetical computation available in electrical investigations. From the many methods of electrical measurement, as applied to the solution of practical problems, a selection has been made, embracing only those which have proved to be most directly applicable to every-day work.

The numerous authorities which have been consulted in the preparation of the present treatise have been carefully indicated in the foot-notes; in many instances with the addition of the titles of publications which may profitably be consulted by the student desiring to investigate more minutely some particular branch of the subject. These references are intended to constitute, in some sense, a key to the standard literature of electricity, although from the nature of the case, by no means an exhaustive one.

It is hoped that the brief biographical notices of men who have distinguished themselves in connection with electrical science will be found to add something to the value of the work, especially as the facts given are sometimes difficult of access to the ordinary reader.

The author acknowledges with pleasure his indebtedness to many friends for courtesies extended, especially to Professor Moses G. Farmer of Eliot, Me., and Messrs. E. M. Barton, of the Western Electric Company of Chicago, E. S. Greeley of New York, and Edward Weston, of Newark, N. J.

EDGEWOOD FARM, ELIZABETH, N. J.,
September 1, 1891.

CONTENTS.

CHAPTER I.
INTRODUCTORY.

Fundamental Principles, §§ 1, 2, 3.—Nature of Electricity, 4.—Elements of Electric Telegraph, 5.................................. 1

CHAPTER II.
SOURCES OF ELECTRICITY.

Origin of Electricity, §§ 6, 7.—Voltaic Element, 8.—Description of the Typical Cell, 9, 10.—Phenomena of Cell, 11, 12, 13.—Chemistry of the Voltaic Effect, 14, 15.—Gravity Cell, 16.—Specific Gravity, 17.—Hydrometer, 18.—Charging the Cell, 19, 20, 21.—Copper and Zinc Solutions, 22.—Specific Gravities of Battery Solutions, 23.—Installation of Gravity Cell, 24, 25.—Modified Form of Copper Plate, 26, 27.—Formation of Electric Circuit, 28, 29, 30.—Nomenclature of Electric Circuit, 31.—Chemical Reactions arising in Closed Circuit, 32.—Effect of Continued Action, 33, 34, 35.—Rate of Consumption of Material, 36.—Maintenance of Cell, 37.—Prevention of Evaporation, 38, 39.—Dismantling Cell, 40.—Diffusion of Solution within Cell, 41, 42.—Neutralizing Zinc Solution, 43.—Waste Products of Cell, 44, 45.—Other Forms of Cell, 46, 47.—Lockwood Cell, 48.—Daniell Cell, 49, 50.—Maintenance of Daniell Cell, 51.—Renewal of Daniell Cell, 52.—Intermingling of Solutions, 53, 54.—Choice of Battery Materials, 55, 56.—General Directions for Care of Cells, 57.—Oxide of Copper Cell, 58.—Setting Up and Maintaining Oxide of Copper Cell, 59.—Chemical Reactions of Oxide of Copper Cell, 60.—Grove and Bunsen Cells, 61.—Wasteless Battery Zinc, 61a.. 3

CHAPTER III.
SOURCES OF ELECTRICITY.—(*Continued.*)

Magneto-Electricity, § 62.—Magnetism, 63.—Magnetic Needle, 64.—Phenomena of Magnetic Induction, 65.—Polarity of Magnet, 66.—Horseshoe Magnet and Armature, 67.—Magnetic Spectrum, 68.—Magnetic Field, 69.—Lines of Magnetic Force, 70, 71.—Attraction and Repulsion, 72.—Electric Current Produced by Magnetic Field, 73, 74.—Transformation of Mechanical Power into Electricity and

Heat, 75.—Direction of Induced Current, 76.—Mutual Reactions of Current and Magnet, 77.—Summary of Magneto-Electric Phenomena, 78.—Dynamo-Electric Machine, 79.—Theoretical Dynamo, 80, 81.—Frictional Electricity, 82.—Thermo-Electricity, 83........ 24

CHAPTER IV.

THEORY OF QUANTITATIVE ELECTRICAL MEASUREMENT.

Electric Current, § 85.—Manifestations of Current, 86.—Importance of Quantitative Measurement, 87. — Fundamental Units of Mass, Space, and Time, 88.—Illustration of Absolute System of Measurement, 89.—Derivation of Electrical and Magnetic Units, 90.—C. G. S. Units of Force and Work, 91.—Conservation of Force, 92.—Electric Field of Force, 93.—Relation of Current Force to Mechanical Force, 94.—Galvanoscope and Galvanometer, 95.—Tangent Galvanometer, 96.—Character of Electrical Measurements, 97.—Characteristics Capable of Measurement, 98.—Apparatus for Measurement, 99.—Ammeter, Voltmeter, and Calorimeter, 100......... 35

CHAPTER V.

THE LAWS AND CONDITIONS OF ELECTRICAL ACTION.

Apparatus Required by Student, § 101.—Construction of Tangent Galvanometer, 102, 103, 104, 105.—Construction of Rheostat, 106.—Preparation for Experiments, 107.—Effect of Varying Number of Cells in Series, 108.—Cells in Parallel Series, 109.—Cells in Parallel, 110.—Increasing Length of Conducting Circuit, 111, 112, 113.—Conditions which Determine Quantity of Current, 114.—Resistance, 115.—Conductors and Insulators, 116, 117.—Specific Resistance of Different Metals, 117a.—Conditions Affecting Resistance, 118.—Provisional Theory of Electricity, 119.—Mechanical Analogue of Electrical Action, 120.—Conception of Potential and Electromotive Force, 121.—Practical Electric Units, 122.—Ampère, 123. — Coulomb, 123a. — Volt, 124. — Ohm, 125. — Resistance of Liquids, 126.—Ohm's Law, 127.—Joule's Law, 128, 129.—Experimental Proof of Ohm's Law, 130.—Internal Resistance of Cell, 131.—First Case, 132.—Second Case, 133.—Law of Joint Resistances, 134, 135.—Third Case, 136, 137, 138, 139.—Branch or Derived Circuits, 140.—Electric Potential, 141, 142.—Illustration of Fall of Potential, 143.—Fall of Potential Proportionate to Resistance, 144.—Graphic Illustration of Electric Circuit, 145.—Fall of Potential in Non-homogeneous Circuit, 146.—Electrostatic Capacity, 147.—Farad, 148.—Power, or Rate of Work, 149.—Watt, 150.—Current Induction, 151.—Electrical Dimensions of Voltaic Cell, 152.—E. M. F. and Resistance of Cell, 153.—Quantity and Cost of Materials Consumed in Battery, 154, 155, 156.—Production of Electricity in Proportion to Material Consumed, 157.—Con-

Contents. xi

sumption of Material in Series of Cells, 158.—Electrical Dimensions of Edison-Lalande Cell, 159.—Effect of Temperature upon Resistance of Metallic Conductors, 160, 161.—Effect of Temperature upon Resistance of Liquids, 162.—Effect of Temperature upon Resistance of Daniell Cell, 163, 164............................... 45

CHAPTER VI.

THE LAWS OF ELECTRO-MAGNETISM.

Elements of Electro-Magnet, §§ 166, 167.—Polarity of Electro-Magnet Determined by Direction of Current, 168.—Lines of Force as a Measure of Magnetic Field, 169, 170.—Unit of Magnetism, 171.—Magneto-motive Force, 172.—Effect of Iron in Helix, 173.—Effect of Magnetization upon Soft Iron, 174, 175.—Magnetic Saturation, 176.—Magnetization Proportional to Ampère-turns, 177.—The Magnetic Circuit, 178.—Magnetic Permeability, 179.—Law of Magnetic Circuit, 180.—Determination of Magnetic Reluctance, 181.—Ratio of Attractive Force to Distance, 182.—Construction of Telegraph Magnet, 183.—Theoretical Proportions of Telegraph Magnet, 184.—Effect of Position of Windings, 185.—Helix of Coil, 186.—Relation of Thickness and Length of Wire to Number of Turns, 187.—Dimensions and Resistances of Magnet Wires, 188.—Thickness of Spaces between Turns of Wire, 189.—American Standard Wire Gauge, 190.—British Standard Wire Gauge, 191.—Instruments for Gauging Wire, 192.—Adaptation of Magnets to Working Currents, 193, 194.—Spectrum of Electro-Magnet, 194.—Magnetic Hysteresis, 195.—Induction of a Current upon Itself, 196. —Magnet Cores must not be Hardened, 197.—Effect of Self-Induction and Hysteresis in Telegraph Magnets, 198.—Other Indirect Causes of Retardation in Electro-Magnets, 199.—Electro-Magnet with Polarized Armature, 200, 201.—Combinations of Permanent and Electro-Magnets, 202... 80

CHAPTER VII.

TELEGRAPHIC CIRCUITS.

Telegraphic Circuits, §§ 203, 204, 205.—Open and Closed Circuits, 206. —Drawings of Electric Apparatus, 207.—Conventional Representations of Circuits and Apparatus, 208.—The Earth as an Electrical Conductor, 209.—Ground Connection, 210.—Advantages of the Earth Circuit, 211.—Open Circuit, 212.—Closed Circuit, 213.—American Modification of Closed Circuit, 214.—Comparative Advantages of Different Plans, 215.—Position of Battery in Closed Circuit, 216.—General Considerations respecting Telegraphic Circuits, 217.—Relation of Conductivity to Insulation Resistance, 218. —Effect of Imperfect Insulation, 219.—Working Efficiency of Telegraphic Circuit, 220.—Telegraphic Conductors, 221.—Iron Wires,

Contents.

222.—Office Wires, 223.—Copper Line Wires, 224.—Telegraphic Line Insulators, 225.—Defects of Glass Insulator, 226.—Resistance Influenced by Form of Insulator, 227.—Hard Rubber Insulator, 228.—Paraffin Insulator, 229.—Porcelain Insulator, 230.—Defective Insulation of American Lines, 231, 232, 233.—Distribution of Potentials in Telegraphic Circuits, 234.—Potential in Perfectly Insulated Circuit, 235.—Determination of Potential by Calculation, 236.—Potentials within the Battery, 237, 238, 239, 240.—Potentials in Imperfectly Insulated Circuit, 241.—Effect of Imperfect Insulation upon Flow of Current, 242.—Resistance and Current in Leaky Lines, 243.—Computation of Working Efficiency of Line, 244, 245.—Effect of Position of Fault, 246.—Best Position of Batteries in Circuit, 247.—Intermingling of Currents on Different Lines, 248.—Remedy for Cross-Current, 249, 250.—Value of Poles and Cross-Arms as Insulators, 251.—Tests of Resistance of Cross-Arms, 252.—Tests of Glass Insulators, 253.—Importance of High Working Efficiency, 254.—Best Method of Improving Efficiency, 255, 256... 102

CHAPTER VIII.

EQUIPMENT OF AMERICAN TELEGRAPH LINES.

Apparatus Essential in Telegraphy, § 257.—Construction of Key, 258.—Modifications of Key, 259.—Adjustment of Key, 260.—Sounder, 261.—Short Line Instrument, 262.—Adjustment of Sounder, 263.—Pocket Apparatus, 264.—Box Sounder, 265.—Working by Relay and Local Circuit, 266.—Construction of Relay, 267.—Adjustments of Relay, 268.—Register, 269, 270.—Adjustments of Register, 271.—Causes of Defective Marking, 272.—Ink-Writing Register, 273.—Circuits of American System, 274.—Arrangements of Apparatus at Way-Station, 275.—Connections of Apparatus of Way-Station, 276.—Manipulation of Switchboard, 277.—Testing for Disconnection, 278.—Reporting Result of Test, 279.—Wedge Cut-Out, 280.—Multiple Wire Switchboard, 281.—Multiple Spring-Jack, 282.—Universal Switchboard, 283.—Manipulation of Universal Switchboard, 284.—Arrangement of Apparatus of Terminal Station, 285.—Terminal Switchboard, 286.—Instrument Tables, 287.—Lightning Arrester, 288.—Plate Arrester, 289.—Safety Fuse, 290.—Inspection and Care of Arresters, 291.—Repeater, 292.—Manual and Automatic Repeaters, 293.—Button Repeater, 294.—Wood's Repeater, 295.—Management of Button Repeater, 296.—Milliken Automatic Repeater, 297.—Management of Automatic Repeaters, 298.—Dynamo-Electric Generator, 299.—Characteristics of Dynamo-Current, 300.—Electro-Magnetic Field, 301.—Commutator, 302.—Characteristics of Dynamo, 303.—Dynamo in Potential Series, 304.—Positive and Negative Dynamo Series, 305.—Arrangement of Shunt Coils, 306.—Capacity of Dynamo Generator, 306a.—Multiple Telegraphy, 307.—Differential Electro-Magnet, 308.—Construction of Differential Magnet, 309.—Single-Current Duplex, 310.—Circuits of Single-

Contents. xiii

Current Duplex, 311.—Artificial Line, 312.—Balancing Resistance, 313.—Electrostatic Capacity of Line, 314.—Electrostatic Accumulation upon Insulated Conductor, 315.—Effect of Currents of Charge and Discharge, 316.—Condenser, 317.—Ground and Spark Coils, 318.—Double-Current Duplex, 319.—Quadruplex, 320.—Principle of Diplex, 321, 322.—Operation of Diplex, 323.—Diplex and Contraplex Combined, 324.—Quadruplex worked by Dynamo-Currents, 325.—Distribution of Currents in Quadruplex Apparatus, 326, 327.—Practical Management of Quadruplex, 328.—Adjustment of Apparatus, 329.—Repeaters for Multiple Telegraph Systems, 329a.. 138

CHAPTER IX.
TESTING TELEGRAPH LINES.

Object of Tests, § 330.—Faults and Interruptions, 331.—Testing for Disconnection, 332, 333.—Testing for Partial Disconnection, 334.—Testing for Escape, 335.—Testing for Cross, 336.—Principle of Cross Test, 337, 338, 339.—Testing by Quantitative Measurement, 340.—Wheatstone Bridge, 341.—Best Ratio of Electromotive Forces and Resistances, 342.—Principle of Wheatstone Bridge, 343.—Actual Construction of Bridge, 344.—Galvanometer for Wheatstone Bridge, 345.—To Measure the Conductivity Resistance of a Telegraph Line, 346.—Conductivity Resistance by Loop Method, 346a. —Earth Currents, 347.—Measurement of Resistance of Ground Plate at Distant Station, 348.—Measurement of Insulation Resistance of Line, 349.—Location of Position of a Ground, 350.—Location of Position of an Escape, 351.—Method of Double Measurement, 352.—Loop Test, 353.—Varley's Loop Test, 354.—To Locate a Cross, 355, 356, 357.—To Locate a Bad Joint or Abnormal Resistance, 358.—Measurement of very High Resistances, 359.—Shunts of Galvanometers, 360.—Measurement by Deflections, 361.—Measurement of Resistance of Insulators, 362.—Measurement of Internal Resistance of Battery, 363, 364.—Measurement of Resistance of Galvanometer, 365.—Differential Galvanometer, 366.—Testing for Insulation by Received Currents, 367.—Use of Voltmeter and Ammeter in Telegraphic Testing, 368.—The Weston Ammeter and Voltmeter, 369.—Recording Tests of Conductivity and Insulation, 370.. 190

CHAPTER X.
HINTS TO LEARNERS.

Formation of the Telegraphic Code, § 371.—The American Morse Code, 372.—Learning the Code, 373.—Handling the Key, 374.—Elementary Principles of Code, 375.—Preliminary Practice with the Key, 376.—Exercises upon Code Characters, 377.—Reading by Sound, 378, 379.—A Parting Word, 380............................... 216

LIST OF TABLES.

		PAGE
I.	Chemical Atomic Weights of Elementary Substances of Batteries........	8
II.	Specific Gravities of Battery Solutions.................	9
III.	Natural Tangents for every half degree...............	55
	Conductors and Insulators in Relative Order...........	57
IV.	Specific Resistances of Voltaic Solutions...............	63
V.	Reciprocals of Numbers from 1 to 100..................	67
VI.	Synopsis of Practical Units............................	73
VII.	Chemical Equivalents of Battery Materials.............	75
VIII.	Dimensions and Properties of Copper Magnet Wires.....	94
IX.	Size, Weight, and Resistance of Telegraph Wires........	112
X.	Resistances and Escape upon Leaky Lines of Various Lengths..	129
XI.	Farmer's Table for Computing Flow of Current in Leaky Lines...	137
	The Morse Telegraphic Code...........................	218

MODERN PRACTICE

OF THE

ELECTRIC TELEGRAPH.

CHAPTER I.

INTRODUCTORY.

1. **Fundamental Principles.**—The electric telegraph is an apparatus by means of which physical effects may be instantaneously produced in distant places. Such effects are technically termed *signals*.

2. The art of *electric telegraphy* consists in the production, control, and organization of electric signals. The signals employed in telegraphy may be either *visible* or *audible*. Visible signals may be either *evanescent*, as in the needle telegraph used in Great Britain, and in some forms of apparatus employed in working long submarine cables, or *permanent*, as in the case of the Morse register, and in the instruments used on submarine cables of comparatively moderate length. Audible signals are produced by the *sounder*, and some other less common forms of apparatus.

3. The signals which are utilized in electric telegraphy are produced at a distant point as required by the agency of *electricity*.

4. **Nature of Electricity.**—We do not as yet know ; perhaps we never shall know with certainty, what the agent we call electricity really is. Formerly it was assumed to be an imponderable fluid. This hypothesis was suggested by Franklin. In later years it gradually came to be regarded as one of the many different forms of energy, or, in other words, as a peculiar affection of the particles of ordinary matter. Recent scientific opinion shows a marked tendency

toward the acceptance of the old hypothesis of a fluid, in a modified form.[1]

This conception of the essential nature of electricity appears to be the logical outgrowth of the opinions held, more or less definitely, by such philosophers as Franklin, Cavendish, Faraday, Henry, Thomson, and, more especially, Clerk-Maxwell. The theoretical side of the question is discussed with great ability by Oliver J. Lodge, in his *Modern Views of Electricity*, while its latest aspects are summarized in a valuable paper by Professor William A. Anthony: "A Review of Modern Electrical Theories," *Electrical Engineer*, ix. 43; *Trans. Am. Inst. Elec. Eng.*, vii. 33.

For practical purposes, however, it is fortunately not in the least necessary that we should know what electricity is, nor that we should commit ourselves to any particular assumption as to its essential nature. A thorough knowledge of the physical effects which it is capable of producing under different conditions, and of the laws which govern its action, are all that the practical electrician needs to acquire.

5. **Elements of the Electric Telegraph.**—An electric telegraph comprises four essential elements. These are as follows:

(i) Means for setting in action, or, as it is commonly termed, producing, electricity, termed the *generator*.
(ii) Means for conducting the electricity from place to place, termed the *conductor* or *conducting circuit*.
(iii) Means for controlling the flow of electricity for the purpose of producing signals, termed the *transmitter*.
(iv) Means for indicating or recording signals, termed the *receiver*.

[1] Electricity and magnetism are not forms of energy; neither are they forms of matter. They may, perhaps, be provisionally defined as properties or conditions of matter; but whether this matter be the ordinary matter, or whether it be, on the other hand, that all-pervading ether by which ordinary matter is everywhere surrounded, is a question which has been under discussion, and which may now be fairly held to be settled in favor of the latter view.—DANIELL: *Principles of Physics* [2d ed.], 532.

CHAPTER II.

SOURCES OF ELECTRICITY.

6. Origin of Electricity.—Electricity, or, more properly, electrical action, may be produced in several known ways. Although electricity, whatever may be its origin, is demonstrably one and the same thing,[1] it has nevertheless become customary to speak of it conventionally, under different names, indicative of its origin. Thus, we have, principally:

(a) CHEMICAL ELECTRICITY.
(b) MAGNETO-ELECTRICITY.
(c) FRICTIONAL ELECTRICITY.
(d) THERMO-ELECTRICITY.

There are other means of producing electrical manifestations, which have as yet no practical utility for the purpose under consideration, and need not be further considered here. Of those which have been specifically mentioned, *chemical* and *magneto-electricity* only, have proved by experience to be adapted to the requirements of the art of telegraphy.

7. Chemical Electricity.—The effects of electricity are most conveniently studied in connection with that form which has its origin in chemical decomposition, especially as it is by the agency of chemical electricity that nearly all the telegraphic apparatus of the world is operated.

8. The Voltaic Element.—The electricity which is employed in telegraphy is usually derived from one or more *batteries*, each of which is composed of a greater or less number of *cells* connected together in a series. A single cell is termed a *voltaic* or *galvanic element*.

9. Description of the Typical Cell.—The active or actuating parts of each element consist in practice of two dissimilar metals, each of which is immersed in a different chemical solution.

[1] This fact was first experimentally established by Faraday in 1832. His account of this investigation is very instructive, and is given at length in his *Experimental Researches* (third series), vol. i. pp. 76-109.

3

Sources of Electricity.

The following will serve to explain the construction most usually employed in telegraphy:

In Fig. 1 is represented a cylindrical glass *jar*, 7 in. in height, 6 in. in diameter, and weighing about 2.5 lbs.

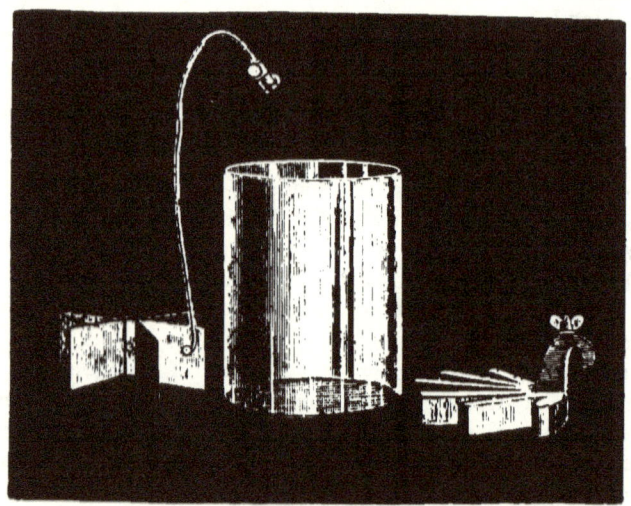

FIG. 1. Separate parts of Gravity Cell.

At the right of this is a mass of *zinc* weighing about 3 lbs., which has been cast in an iron mould in the form represented. It is provided with a *hanger* by means of which it may be suspended from the upper edge of the jar, and also with a *clamp-screw* by means of which a metallic wire may be securely, but removably, attached to it.

At the left of the jar is seen a triple plate of thin rolled *copper,* spread out laterally into the form shown, which is designed to be placed in the bottom of the glass cell. Each separate plate may be cut in the form shown in Fig. 2, the three being then united by a single copper rivet at the middle, and the free ends separated radially, as in Fig. 1, before placing in the jar.

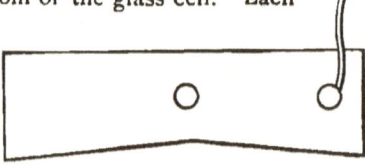

FIG. 2. Section of Copper Plate.

A vertical copper *wire* is permanently riveted to the copper plate. It must not be soldered. The wire is made long enough to extend some 6 in. or 7 in. above the top of the jar. It passes loosely through the bore of a small glass tube, the reason for which is hereafter explained (27). It is provided at its upper end with a brass clamp termed the *copper-connector*. Instead of using a glass tube, it is quite usual to substitute a piece of wire covered with a coating of gutta-percha, india-rubber, or other flexible material impervious to the solution.

The particular form of copper-connector shown in the figure consists of a short cylindrical piece of brass, perforated with a longitudinal hole for receiving the ends of the wires, into which enter transverse thumb-screws for clamping the wire. A longitudinal cross-section of this device is shown in Fig. 3.

FIG. 3. Copper-connector.

10. Each element, when complete, consists of the several parts described, assembled together in the relation shown in Fig. 4, which also shows the jar filled with water to within 1 in. of the top. Care must be observed, in hanging the zinc, not to fracture the glass jar.

It is very essential that the water for charging a voltaic element should be both pure and soft. Impure or hard water obstructs, and sometimes altogether prevents, the proper action of the chemicals. Clean rain-water, if it can be procured, is best for the purpose.

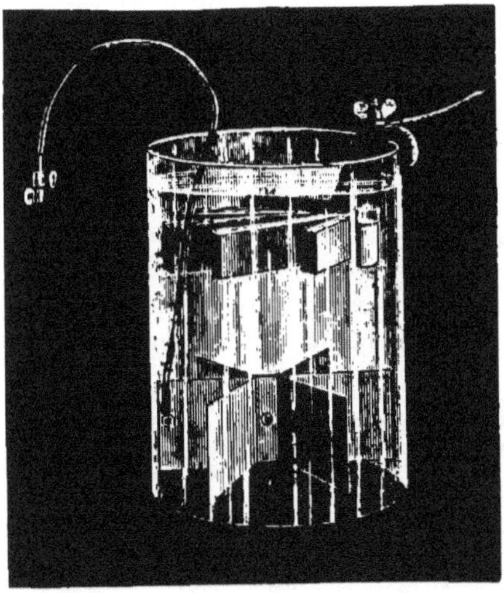

11. **Phenomena of the Cell.**—If a cell, having been thus filled with water, in which zinc and copper plates are immersed, as shown in Fig. 4, be permitted to stand undisturbed for a considerable time, a collection of minute bubbles will be observed clinging to the surface of the zinc plate, but no such effect will be observed upon the copper. These bubbles contain *hydrogen* gas, and are the result of a chemical reaction which takes place between the water and the zinc.

FIG. 4. Gravity-cell ready for Service.

12. Water is made up of two parts of hydrogen and one part of oxygen, held together by chemical affinity. In the present case, a

certain portion of the oxygen of the water enters into chemical combination with the metal, forming the compound termed *oxide of zinc*. A thin coating of this oxide ultimately covers the surface of the zinc plate, giving it a dull bluish gray color, and preventing further oxidization. At the same time the hydrogen which was associated with the oxygen in the decomposed water, is set free, and collects in bubbles which adhere to the surface of the zinc plate. When these bubbles are detached they rise to the surface of the water and the contained hydrogen escapes into the air.[2]

13. This process of oxidization will be recognized as identical with that which takes place in the rusting of iron when exposed to the action of moisture. It is also the same, from a chemical point of view, as the process of *combustion* or burning. No perceptible effect is produced upon the copper, because the oxygen has less affinity for this metal than it has for hydrogen, and hence has no tendency to separate from the water.

14. **Chemistry of the Voltaic Effect.**—If now a small quantity of sulphuric acid were to be added to the water contained in the jar, and at the same time the zinc and copper be joined together by a metallic wire in the air outside the jar, a much more vigorous chemical action will immediately set in. The dissolution of the zinc in the liquid will go on with increased rapidity, attended with the evolution of hydrogen in bubbles, not as before upon the surface of the zinc, but upon the copper.[3]

15. Although the chemical action in the case just supposed is attended by the development of electricity, yet such an organization, as a generator of electricity for telegraphic purposes, would be of little practical utility. The chemical action, though vigorous at first, quickly falls off, and in a short time nearly or quite ceases. This effect arises from the adherence of the liberated hydrogen to the sur-

---

[2] In a chemical compound the qualities of the constituents are wholly merged in those of the product, and this circumstance distinguishes a true compound from a mechanical mixture in which the qualities of each ingredient are to a greater or less extent preserved. . . . Chemical combinations always take place in certain definite proportions, either by weight or measure. . . . The atomic theory supposes that two atoms of hydrogen combine with one atom of oxygen to form a molecule of water, and since each atom of oxygen weighs 16 times as much as an atom of hydrogen, the two substances must combine in the proportion of 2 : 16, or 1 : 8. This principle is known in chemistry as the *law of definite proportion.*—COOKE : *New Chemistry*, 104–8.

[3] The chemical reaction is as follows :—Sulphuric acid is composed of *hydrogen* 2 parts, *sulphur* 1 part, *oxygen* 4 parts; in chemical notation (H_2SO_4). The sulphur and oxygen unite with the zinc, forming *sulphate of zinc*, composed of zinc 1 part, sulphur 1 part, and oxygen 4 parts ($ZnSO_4$), which remains in solution in the water, while the hydrogen is set free at the copper plate.

face of the copper, preventing contact of the solution therewith. This gas also reacts upon the sulphate of zinc (*s. z.*) which permeates the solution, and causes its zinc constituent to be deposited upon the copper. For these reasons it is necessary to dispose of the hydrogen in such a way that interfering actions may be avoided. This is effected in practice by immersing the copper and zinc in different solutions.

16. **The Gravity Cell.**—In the voltaic element which has been described and shown in Fig. 4, the two solutions are of unequal densities, so that one can be made to float, as it were, upon the other, in the same manner that oil floats upon water. Hence it has received the name of the *gravity cell.*

17. **Specific Gravity.**—The density or weight of a given bulk of any liquid compared with that of pure water is termed its *specific gravity* (*s. g.*) The *s. g.* of a liquid is numerically expressed in decimals or mixed numbers, pure water being taken as the standard or unity. For example, the *s. g.* of water being 1.00, that of linseed oil is 0.93, while that of commercial sulphuric acid is 1.84, and of mercury 13.58.

18. **The Hydrometer.**—The *s. g.* of any liquid may be determined with sufficient accuracy for ordinary purposes by means of the *hydrometer,* shown in Fig. 5, which consists of a hollow glass float, weighted below with shot, and carrying a stem at the top provided with a graduated scale. When the hydrometer is made to float in any liquid, the division of the scale at the surface denotes its *s. g.*[4]

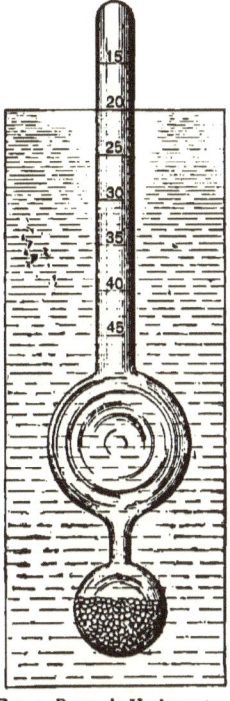

FIG. 5. Baume's Hydrometer.

19. **Charging the Cell.**—The glass jar shown in Fig. 4, which

---

[4] The arbitrary scale of the hydrometer commonly known as Baume's, is determined as follows:—The point to which the instrument sinks in pure water is assumed as 0° (zero), while 15° is at the point to which it sinks in a solution containing 15 parts by weight of common salt in 85 parts of water. This space is divided into 15 equal parts, and equivalent graduations are continued to any desired extent. The most useful scale for testing the *s. g.* of battery solutions is one having a stem about 2 inches long, graduated in degrees from 15° to 40°. These degrees denote the percentage of common salt in a solution; but do not correspond exactly with the percentages in battery solutions, as will appear from an examination of the tables in (23).

is 7 in. high and 6 in. in diameter, is intended to contain 7 lbs., or 0.84 U. S. gallons of liquid ; and this quantity, when the copper and zinc plates are in place, will fill it to within about one-half inch of the top. A smaller size of cell (6 in. × 5 in.), is also kept in stock by dealers.

20. To charge the cell, prepare separately a sufficient quantity of the zinc and of the copper solutions. For the *zinc solution*, which may be mixed in the jar, take for each cell:

> Pure soft water, by weight, 91 oz. (1½ pints).
> Crystallized sulphate of zinc (white vitriol), 10 oz.

Dissolve, and let the solution stand for some hours. The *s. g.* of the solution should be 1.10.

21. It is not absolutely necessary to make use of *s. z.* in setting up the cell. If pure water be substituted for the solution directed to be used in the last paragraph, and the circuit be closed between its poles, which is technically termed *short-circuiting* the cell (14), sufficient *s. z.* will be formed within a day or two to bring it into full action. Many electricians are of the opinion that a cell started in this way will remain in good condition for a longer time than if charged with a mechanically-mixed zinc solution.

22. **Copper and Zinc Solutions.**—For the copper solution, take in another glass vessel, for each cell:

> Pure soft water, by weight, 42 oz. (2¼ pints).
> Crystallized sulphate of zinc, 4 oz.
> Crystallized sulphate of copper, 8 oz.

The following table will be found convenient for reference:

TABLE I.

CHEMICAL ATOMIC WEIGHTS OF ELEMENTARY SUBSTANCES OF BATTERIES.

SUBSTANCE.	SYMBOL.	ATOMIC WEIGHT.
Copper.............	Cu	63.4
Zinc	Zn	65.2
Sulphur............	S	32.0
Oxygen	O	16.0
Hydrogen	H	1.0

The chemical notation for crystallized sulphate of copper is $(Cu. SO_4 5H^2O)$. It is composed of

> Metallic Copper............................ 25.4 per cent.
> Sulphur................................... 12.8 "
> Oxygen .... 57.7 "
> Hydrogen .... 4.0 "

Specific Gravities of Battery Solutions. 9

The chemical notation of crystallized sulphate of zinc is (Zn SO⁴ 7H²O). It is composed of

Metallic Zinc	22.7 per cent.
Sulphur	11.1 "
Oxygen	61.3 "
Hydrogen	4.9 "

When dissolved, the *s. c.* solution will be of a beautiful dark blue tint, and its *s. g.* will be 1.21.

23. Specific Gravities of Battery Solutions.—The following tables will aid in maintaining cells is good condition:

TABLE II.
SPECIFIC GRAVITIES OF BATTERY SOLUTIONS.
ZINC.

S. g. of solution at 77° Fah.	Reading by Baume hydrometer.	Per cent. of crystallized s. z. in solution.	REMARKS.
1.11	15	15.7	Minimum density.
1.12	16	16.8	
1.13	17	17.9	
1.135	18	18.9	
1.14	19	20.0	
1.15	20	21.1	Maximum density.
1.16	21	22.3	
1.17	22	23.4	
1.18	23	24.6	
1.19	24	25.8	
1.20	25	26.9	
1.46	48	62.1	Saturation.

COPPER.

S. g. of solution at 72° Fah.	Reading by Baume hydrometer.	Per cent. of crystallized s. c. in solution.	REMARKS.
1.03	5	5.0	
1.07	10	10.0	
1.11	15	15.4	Half saturation.
1.15	20	21.2	
1.20	25	27.5	
1.21	27	30.0	Saturation.

24. Installation of the Gravity Cell.—The copper and zinc plates being put in their respective places in the jar (which will then be about three-fourths full of *s. z.* solution), the heavier *s. c.* solution may be introduced into the bottom by means of a ⅜ in. tube of glass or rubber, having a small glass or rubber funnel inserted in its upper end. The lower end of the tube must be central and very near the bottom, and the *s. c.* must be poured in quite slowly, so as not to agitate the mass and cause the two solutions to mingle. If this operation is carefully performed, the lower part of the jar will now be filled with *s. c.* solution, of a uniform deep blue color, to a point a little above the top of the copper plate, being separated from the transparent *s. z.* solution above by a sharply defined line of demarkation. Care must now be taken that the cell is not moved about, shaken, or stirred by the careless removal of the zinc or copper plates, as this would cause the two solutions to intermingle, a condition which it is very necessary to avoid. For the same reason, it is advisable to place each cell in the position which it is to permanently occupy, before introducing the *s. c.* solution. The most convenient place will be found to be a shelf about 48 in. from the floor. An enclosed box affixed to a wall or frame, and having a glass front hinged to open upward, is an excellent arrangement, as the cells are then in sight, so that their condition may be observed at all times, while at the same time they are protected from dirt, and in a great measure from evaporation and from extremes of temperature.[5]

25. Instead of mixing the *s. c.* solution in a separate vessel, it is a common practice to fill the jar to within 1 inch of the top with the *s. z.* solution, prepared as above directed, and then slowly drop in 8 oz. of *s. c.* crystals about the size of a hazel-nut, which will fall to the bottom and slowly dissolve. The only objection to this procedure is its liability to form a *s. c.* solution of unequal density in different parts, which is undesirable (26). When this plan is adopted, care must be taken not to put in *more* than the prescribed quantity of *s. c.*, and particularly to see that no particle of it gets upon the zinc plate.

26. **Modified Form of the Copper Plate.**—A more advantageous form for the copper plate than that which has been described,

---

[5] Wooden, tin, or porcelain covers are sometimes fitted to the cells for excluding dust and preventing evaporation, and serve a good purpose. Wooden covers should not fit too closely; there is danger that they may swell from moisture and fracture the jars.

Great annoyance is sometimes caused by the apparently unaccountable breakage of glass jars. The primary cause of this is poor material or imperfect annealing during the process of manufacture; the immediate cause is usually a sudden change of temperature. A jar on a high shelf in a warm room in winter is sometimes cracked by the current of cold air caused by opening an outer door. A little care will avoid such accidents.

particularly in case it is desired to maintain a current of moderate quantity for a long time, is a ribbon of very thin rolled copper, 48 in. long and $\frac{1}{2}$ in. wide, coiled spirally like a clock-spring, and laid flat in the bottom of the cell, the conducting wire being riveted to the outer end as seen in Fig. 6. An objection to the form of plate shown in Fig. 4, when used under the conditions here mentioned, is that unless carefully looked after, the *s. c.* solution will become weaker at the top than at the bottom of the copper, whereupon a closed circuit (3c) is established, consisting of one metal (the copper), and two dissimilar liquids (the strong and the weak solution), setting up an action which is liable to attack and destroy the upper portion of the plate, uselessly consuming material for which no equivalent external current is rendered.

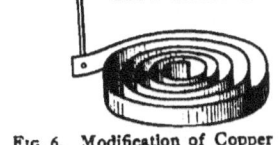

FIG. 6. Modification of Copper Plate.

27. This action explains the necessity of enclosing the connecting wire from the copper electrode of the gravity battery in a glass tube, or covering it with gutta-percha or india-rubber, where it is exposed to the action of the solution. If it were not protected it would soon be destroyed by chemical action, and the circuit consequently interrupted.

28. **Formation of the Electric Circuit.**—The parts of the cell being properly assembled together, and the solutions in their respective places as directed in (24), the element is ready for service. If now the zinc and copper plates be joined together in the air by a metallic wire as before explained (14), a *current of electricity*, as it is technically termed, will traverse the wire. It will traverse, moreover, not only the wire, but also the metallic plates and solutions within the voltaic element, the whole path forming what is termed a circuit of electrical conductors, or briefly, an *electric circuit*.

29. The presence of an electric current in such a circuit may be demonstrated in several different ways, as will be shown further on (86). For the present we are only concerned to observe its immediate effects upon the constituent parts of the voltaic element which sets it in action.

30. The circuit of a voltaic element may be diagrammatically represented by a closed ring as shown in Fig 7. It is composed of the following parts :—

(1.) The zinc plate.
(2.) The zinc solution.
(3.) The copper solution.
(4.) The copper plate.
(5.) The metallic connecting wire.

The four first named constitute the *internal circuit*, and the last the *external circuit*.

Before the zinc and copper plates are united by the connecting wire, the circuit is said to be *open* or *broken*, and the cell is said to be on *open circuit*. When the connection is established by the wire, the circuit is said to be *made*, *completed*, or *closed*, the last mentioned phrase being most usual. In this case the cell is spoken of as being on *closed circuit*, which is another way of saying that chemical action is going on within it.

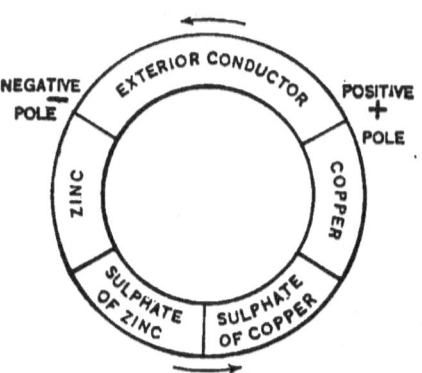

FIG. 7. Diagram of Closed Voltaic Circuit.

31. Nomenclature of the Electric Circuit.— In a voltaic cell, the zinc plate is termed the *positive* plate or element, and the copper the *negative* plate or element. These terms are purely conventional and arbitrary, and properly signify nothing beyond the antagonistic or opposite electrical condition which exists. The general term for these plates is *electrodes*, a term introduced by Faraday. The air terminals of the electrodes, to which the conducting wires are attached, are called the *poles*. It should be noted that the *copper* plate of the element, although the *negative electrode*, is connected with the *positive pole*, and in like manner the *zinc* or *positive electrode* is connected with the *negative pole*, because the current is conventionally assumed to flow from the positive plate, through the solution and out by the copper plate. The positive and negative poles of every generator of electricity are respectively designated by the conventional signs + and − (plus and minus). The direction of the electric current, for convenience of description, is *conventionally assumed*, as above stated, to be through the solutions from the zinc to the copper electrode, and thence through the connecting wire from the copper to the zinc electrode. The assumed direction in any wire is denoted by the conventional sign of an arrow pointing in the direction of the negative pole.

32. Chemical Reactions Arising in the Closed Circuit.— The chemical reactions within the cell, when its external circuit is closed, and its several constituent parts traversed by an electric current, are as follows:

(1.) The oxygen of the *s. z.* solution (12) combines particle by particle with the metal of the zinc plate, forming oxide of zinc.

(2.) The oxide of zinc, formed as above, combines with the sulphuric acid of the *s. z.* solution (14), and forms *s. z.*, which is added to the *s. z.* already present in the solution surrounding the zinc.

(3.) Oxygen combines with the *s. c.* and forms oxide of copper.

(4.) The copper in this oxide separates from the oxygen and is deposited in a pure metallic form upon the copper plate.

At the surface of the zinc plate, the oxygen of the water contained in the *s. z.* solution is separated from the hydrogen, while at the surface of the copper plate this hydrogen combines with the oxygen which is separated from the oxide of copper.

33. **Effect of Continued Action.**—This action goes on without cessation, provided the circuit remains closed, until some one of the materials contained in the cell becomes exhausted. It will be observed that as the action continues, the zinc plate is gradually dissolved, being oxidized, or in fact burned; that the proportion of sulphate in the *s. z.* solution constantly increases, rendering it more dense and its *s.g.* greater; that, on the contrary, the *s. c.* solution grows less dense, and its *s.g.* diminishes ; and finally, that the copper plate continually increases in weight, by the deposition upon its surface of metallic copper abstracted from the copper solution in which it is immersed.

34. As the weight of the *s. z.* solution, as indicated by its *s.g.*, gradually increases, while on the contrary that of the *s. c.* solution continually becomes less, it necessarily happens after the lapse of a greater or less time, the former becomes heaviest, and consequently descends to the bottom of the cell, forcing the *s. c.* solution to the top, where it is brought into direct contact with the zinc plate, depositing metallic copper thereon. This deposit interrupts the normal chemical action of the cell to such an extent that the electric current greatly diminishes, and ultimately ceases altogether.

35. By intelligent management this injurious action may be prevented, or at least postponed for a long time. The frequent use of the hydrometer.(18) is almost indispensable for this work, and a knowledge of the condition of the cell is also greatly facilitated by placing it in front of a window, so that its interior may be clearly viewed by transmitted light ; or at all events, it should be provided, if possible, with a white background.

36. **Rate of Consumption of Material.**—The rapidity with which the materials of the cell are consumed, and its active life shortened, depends almost entirely upon the amount of work done by it,

or in other words, the quantity of electricity per unit of time which it is required to furnish. This question, which is an exceedingly important one, will be fully considered further on (154).

37. Maintenance of the Cell.—The first sign of the exhaustion of a cell generally appears in the $s.\ c.$ solution. It is not practicable to examine the condition of this solution by means of the hydrometer. but fortunately the degree of intensity of its blue color furnishes an infallible indication of its density. The strong blue tint of the original solution will after a time begin to fade in the vicinity of the upper edge of the copper plate, and the line of demarkation between it and the zinc solution will become less and less distinct. When this is seen to occur, additional $s.\ c.$ must be supplied, either through the tube in the form of a solution as directed heretofore (24), or by dropping 1 oz. of crystals into the jar, being careful to observe the precautions heretofore noted (25). It is much better not to make use of finely powdered $s.\ c.$ for this purpose, as this is liable to cement itself into a hard insoluble mass at the bottom of the cell, which defies all efforts to remove it without breaking the jar. The $s.\ z.$ solution should be tested by means of the hydrometer (18) at least once a week while the cell is in constant action. The $s.\ g.$ of the solution, which at the outset was about 1.10, will gradually increase. When it reaches 1.15, as shown by the scale, the solution should be diluted with water. If the $s.\ z.$ solution be permitted to approach closely to its saturation point, 1.45, see table (23), not only is the chemical action of the cell diminished, but a saline deposit of white powder (crystallized sulphate of zinc) begins to form upon the zinc, and upon the edge of the jar above the solution, and by capillary attraction ultimately conveys the liquid over the edge to the outside of the cell and creates a disagreeable nuisance. This may be avoided by keeping the $s.\ g.$ of the zinc solution below 1.20 and by occasionally wiping the inner edges of the jar with a cloth or sponge saturated with cotton-seed or heavy paraffin oil.

38. Prevention of Evaporation.—Sometimes a thin stratum of one of the oils above mentioned is gently poured upon the top of the zinc solution, after the cell has been set up as directed in (24), a procedure which effectually prevents evaporation and the formation of saline salts. Inasmuch, however, as the presence of the oil renders the cleaning of the zinc plate, when necessary, a disagreeable and inconvenient task, it is perhaps an open question whether the practice is to be recommended. If it is at all possible to give the cell proper attention from time to time as required, it is probably better to dispense with all such expedients, but when such is not

Prevention of Evaporation. 15

the case, it may be advisable and even necessary to make use of them.[6]

39. The best way to dilute the zinc solution is to use a tube of rubber, glass, or lead, about 24 in. long, and $\frac{1}{2}$ in. diameter, bent into a siphon, or an inverted **U**, one leg of which is considerably longer than the other. Fill the siphon with water, stopping both ends with the fingers, and after placing a wooden bucket or other convenient receptacle in front of the cell, but at a considerable lower level, dexterously insert the tube into the cell (at the same time removing one finger), so that the inserted end will be near the center of the jar and about $\frac{1}{4}$ in. above the copper plate, while the longer end is directed toward the bucket. Now withdraw the other finger from the lower end of the tube, and the solution will flow in a steady stream into the bucket so long as the short end of the tube remains immersed (See Fig. 8). Some prefer, instead of a siphon, to use a large syringe, sold by dealers, with a nozzle at right angles to the barrel, having a capacity of about 3 gills. This should be rinsed out in warm water each time after it has been used.

After withdrawing about 1 quart of the solution in this way, which with the cell under consideration will be about 2 in. of vertical depth, refill the cell to its original depth, $\frac{1}{2}$ in. from the top, with pure soft water. In order not to stir up the liquids, this may with advantage be done with a small sprinkling pot having a fine rose at the end of its spout, or with due care, may be equally well effected by holding a spoon or some such implement just at the surface, so as to break and scatter the vertical force of the stream as it is poured.

FIG. 8. Drawing off Zinc Solution.

[6] Another device which is sometimes resorted to for the purpose of preventing the formation of salts upon the edge of the jar, is to invert the latter before using, and dip it in a bath of melted paraffin contained in a shallow dish, to the depth of half an inch or less, which forms, when cold, an adherent and repellent coating.

40. Dismantling the Cell.—The above described operation, if properly carried out, will practically restore the cell to its original working condition. The increasing deposit upon the copper plate will not interfere with the proper action of the cell, and need not be disturbed. The zinc plate, however, will gradually become covered with a thick coating of dark brown oxide, which will adhere to it with considerable tenacity. This must be removed from time to time, especially when, by becoming of a reddish color, it shows traces of deposited copper. Lift the zinc plate carefully from the solution, and remove the crust which has formed upon the metal, by means of a scraper of hard wood, or a stiff brush sold by dealers in supplies for that purpose (a wire brush answers the purpose admirably). Remove all the oxide down to the surface of the metal, wash the latter in clean water, and return to its place in the cell. If any undissolved crystals of $s. c.$ are found in the bottom of the jar, these should be washed and used again.

The zinc should be cleaned at once after removal from the cell, while still wet. If the cleaning has to be deferred, the zinc must be placed in water for some time before commencing operations. Great care must be taken to see that no water gets between the arm of the zinc and the brass binding-screw, as this will cause a deposit of sulphate of zinc, which may entirely prevent the passage of the current when the zinc is again put to use.

41. Diffusion of Solution within the Cell.—An absolute separation of the copper and zinc solutions in the voltaic cell cannot be attained. Liquids of unlike density separated from each other by gravity always tend to intermingle by a slow process of diffusion, and thus ultimately to form one homogeneous solution. This tendency may be reduced to a minimum by intelligent management and proper attention to the requirements of the cell while in action, so as to cause but little practical inconvenience.

42. The solutions manifest a much stronger tendency to mix when the cell is on open than when on closed circuit. Hence, cells in which the solutions are separated by gravity, and in fact all sulphate of copper cells, give the most satisfactory results when used, as in telegraphy, upon circuits which are closed the greater portion of the time.

43. Neutralizing the Zinc Solution.—When the cell is dismounted and renewed, the $s. z.$ should be drawn off with the siphon and thrown into a wooden vessel, together with a few pieces of metallic zinc, which will purify the liquid by reducing any metallic copper which may be present in it. It should then be filtered or strained through cloth or sand, and afterward diluted with water until its specific gravity is reduced to 1.10. It is then in suitable condition to be used in the renewed cell, instead of making a new solution as directed in (20).

44. **Waste Products of the Cell.**—Where a large number of cells are in constant use, it is generally worth while to dry and preserve the material thus removed from the zincs, commonly called "battery mud," as it is rich in metallic zinc and copper, and will usually be willingly purchased at a fair price by brass-founders. When the copper plates have become heavily encrusted with metallic deposits, they may with advantage be disposed of in the same way. Electrotype or deposited copper, as this is termed, is much valued in many of the industrial arts.

45. Copper plates which have been used in the battery, and which are intended to be used again, should be kept in water; taking care that the connecting wire, with its coating of gutta-percha or india-rubber, is completely immersed. Zinc plates, on the contrary, must be kept in a dry place, never in water.

46. **Other Forms of the Cell.**—Much unprofitable ingenuity has been displayed by inventors in varying the form, proportions and relations of the elements of the sulphate-of-copper cell, in pursuit of imaginary advantages. As a matter of fact, it has been found to be almost wholly immaterial what the form and arrangement of the parts may be, so long as the necessary general principles of action are kept in view. *The consumption of a given amount of zinc and sulphate of copper can never in any chemical combination, or under any circumstances, evolve more than a definite and perfectly well ascertained quantity of electricity, in a form available for use,* although if the cell be unskilfully proportioned or arranged, the quantity of electricity evolved may be less than it should be (154). The principal difference between different forms is that some require less frequent attention than others; but this advantage is sometimes gained at the expense of other more valuable qualities.

47. Among the different practical voltaic cells which have been employed in America to a greater or less extent, commonly known by the names of their originators and designers, but involving essentially the same principles as the one which has been described, may be mentioned the Hill,[7] Callaud,[8] Minotto,[9] Thom-

---

[7] L. BRADLEY in *The Telegrapher*, iii. 153; E. A. HILL in *the same*, iii. 201.

[8] BLAVIER: *Telegraphie Electrique*, i. 271; POPE: *Modern Practice of the Electric Telegraph* (4th ed.), 106.

[9] F. JENKIN: *Electricity and Magnetism*, 225.

son,[10] etc., etc., for a particular description of which recourse may be had to the publications indicated in the references.

48. **The Lockwood Cell.**—This form of cell has been found to give excellent results in cases in which a moderate but perfectly uniform current is required without attention for a great length of time. The jar is of extra depth (9 in.) and the copper plate consists of two flat spirals of wire coiled like a clock-bell and laid in reverse directions to each other, one beneath and the other at the top of a mass of 5 lbs. of $s.\ c.$ in crystals, placed in the bottom of the jar. The connecting wire is continuous with the lower spiral, while the two spirals are united by a vertical rod or stout wire which is connected to their inner ends. The action of the current traversing the coils appears to act, in some manner not well ascertained, to oppose the tendency of the $s.\ c.$ solution to ascend in the jar and reach the zinc plate. A series of these cells will maintain a current for a year under favorable conditions.

49. **The Daniell Cell.**—This is the original form of the sulphate of copper element. It was formerly much used in the telegraphic service, but has now been practically superseded by the equally efficient and more economical gravity cell. As usually constructed, the Daniell cell consists of a jar of glass or earthenware F (Fig. 9) 6 in. in diameter and 8 in. high. A thin sheet of copper G is bent into a cylindrical form so as to fit loosely within the jar, and to this is affixed a chamber provided with a perforated bottom, designed to receive a supply of $s.\ c.$ in crystals. A copper strip is riveted to the plate G and provided with a clamp at its extremity, adapted either to receive a conducting wire, or to connect to the zinc plate of the next adjacent element, as the case may be. Within the copper cylinder is a porous-cup (as it is technically termed), H, of unglazed porcelain ware, 7 in. high and 2 in. diameter, within which is placed a bar of cast zinc of the cross-section shown at X, or as sometimes preferred, a hollow cylinder with a vertical slit in one side, the latter form yielding a somewhat greater quantity of electricity, but being less convenient to clean.

50. The porous-cup H is filled with $s.\ z.$ solution prepared as directed in (20) and the jar outside the porous-cup with $s.\ c.$ solution of $s.\ g.\ 1.10$. A quantity of the crystals may be placed in the perforated chamber attached to the copper plate, which gradually dissolve and thus maintain the solution at its proper density. Pure water may be used in the porous cell as directed in (21) if preferred.

[10] F. JENKIN : *Electricity and Magnetism*, 223.

51. **Maintenance of the Daniell Cell.**—This cell is maintained in substantially the same manner as the gravity. Unless a very large volume of current is required, it will be found much more

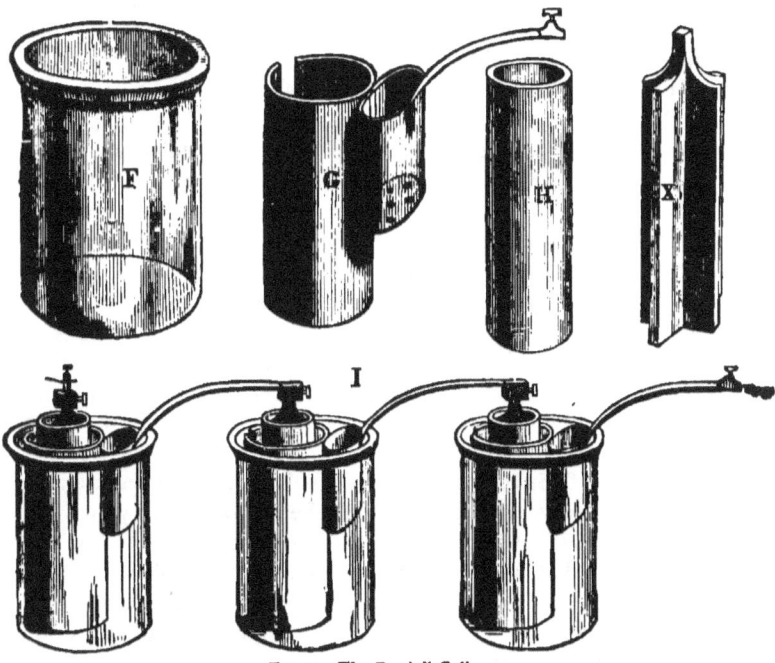

FIG. 9. The Daniell Cell.

economical to feed the *s. c.* solution with small quantities of crystals, placed in the chamber once in every two or three days, and keeping the solution but half saturated (*s. g.* 1.10) and uniform in color throughout, by stirring it with a wooden or glass rod. The *s. z.* solution should be maintained as nearly as possible at the same *s. g.* as the copper solution.

52. **Renewal of the Daniell Cell.**—When taken apart for cleaning, more or less copper will usually be found deposited in patches on the porous-cup. This deposit cannot be prevented, but may be greatly diminished by suspending the zinc free from the bottom or sides of the porous-cup, or even by placing a piece of glass in the bottom of the cup for the zinc to stand on. It is also a good plan, for the same reason, to saturate the bottom of the porous-cup to the height of half an inch with melted paraffin or tallow before putting it to use. The porous-cup ought to be replaced by a new

one whenever as much as half of its surface has become encrusted with metallic copper by continued use. If it becomes cracked it should be replaced at once, or a great waste of material will ensue.

The porous-cup of an element intended only for occasional use, may with advantage be made thicker and less porous in texture than if intended to be kept continuously in action.

53. Intermingling of the Solutions.—It should be observed that at the best, a porous cell merely obstructs and does not prevent the ultimate intermingling of the copper and zinc solutions. The liquids will pass through the porous wall of the cup by virtue of a singular property, common to all dissimilar liquids when separated by a porous partition,[11] and will be found to exhibit a constant tendency to rise in the outer cell and to disappear from the porous-cup. This tendency is obviously assisted by the passage of the current.

54. Porous-cups which have been used in a cell, must not be allowed to become dry after being taken out, but should be kept in water, otherwise the crystallization of the $s.\ z.$ contained in the pores will almost certainly break them.

55. Choice of Battery Materials.—The $s.\ c.$ and the metallic zinc used for electrical purposes should be of good quality and free from adulterations. Adulterated $s.\ c.$ is very seldom met with in the United States; that sold by dealers in electrical supplies is almost uniformly of good quality. The best commercial zinc usually contains a small proportion of iron and lead. An analysis of spelter of good quality for electrical purposes, gave:

```
Zinc.......................................... 98.76 per cent.
Lead .........................................  1.18    "
Iron..........................................  0.06    "
```

56. The question of the effects of temperature upon the efficiency of the voltaic cell is a very important one, and merits much more consideration than it has hitherto received. The sulphate of copper cell is especially sensitive in this particular, and should be carefully guarded against cold. This subject is further considered in a subsequent chapter (162).

57. General Directions for the Care of Cells.—The directions for the management of the sulphate of copper element may be summarized as follows:

(1.) Place the cells in a clean, dry, and well lighted situation, not exposed to dust nor to extremes of temperature.

---

[11] JOHNSON'S *Univ. Cyclopedia*, Art. *Endosmose*.

(2.) Do not move, shake, or stir the cells after the *s. c.* solution has been introduced into them.

(3.) Start each cell with *s. z.* solution at *s. g.* 1.10 (or 15° Baume), and *s. c.* solution not below *s. g.* 1.20 (or 25° Baume).

(4.) Keep the *s. c.* solution of a strong blue color up to a point just above the copper plate, by adding *s. c.* as fast as it is consumed by the action of the current, but be careful never to put in too much *s. c.* at one time.

(5.) Test the *s. z.* solution frequently with the hydrometer, and when its *s. g.* reaches 1.15 (or 20° Baume), dilute with water to reduce it to 1.10 (15° Baume).

(6.) Wipe off with a greasy cloth any crystallized *s. z.* which forms upon the edges of the jars.

(7.) Do not let the zinc become too heavily coated with brown oxides. If the oxides tend to form into pendants, hanging below the zinc, detach these at once with a bent wire; they cause a great waste of material.

(8.) It is an excellent plan to wrap the zinc neatly in linen paper (the kind called parchment paper is best), securing the folded flaps at the top with sealing-wax, and tying strongly with twine passed several times around the whole. This expedient prevents particles of zinc from falling on the copper, and also aids the action of gravity in preventing the too rapid upward diffusion of the *s. c.* solution.

58. **The Oxide of Copper Cell.**
—A voltaic combination in which the metallic elements are amalgamated zinc [12] and protoxide of copper (Cu O), [13] and the exciting agent a solution of caustic potash (K O), has of late found much favor in the telegraphic service, under the name of the Edison-Lalande cell. In the size designed for this use, the glass containing-jar is 8 in. high, 6 in. in diameter, and

FIG. 10. Oxide Plate of Edison-Lalande Cell.

[12] Zinc which has been immersed in dilute sulphuric acid, and then coated with mercury, is said to be *amalgamated*. This process renders the chemical action upon the zinc more uniform and less wasteful in certain forms of voltaic elements. It is of no advantage in the sulphate of copper element.

[13] Protoxide of copper is obtained by roasting copper turnings. The product is then ground to powder and compressed into solid masses, from which are cut plates of suitable size for the cell.

weighs 5.75 lbs. It is provided with a porcelain cover, from which are suspended two rectangular plates of rolled zinc, fitted with a double clamp-screw for attaching the wire. A skeleton frame of copper (Fig. 10) is fitted to clasp two rectangular slabs containing 1 lb. of copper oxide, and is suspended from the porcelain cover between and facing the zinc plates. To prevent possible contact, a fender of hard rubber is inserted between the oxide plates, projecting on each side. A transverse copper bolt and nut clamps the whole firmly together. Fig. 11 shows the appearance of the cell when mounted.

FIG. 11. Edison-Lalande Cell.

59. Setting Up and Maintaining the Oxide of Copper Cell.—The solution for this cell consists of 1 part by weight of caustic potash dissolved in 3 parts pure soft water ($s. g.$ 1.33; 38° Baume), with which the jar is to be filled to within 1 in. of the top. Caustic potash, in sticks of a size just sufficient to make the proper solution, are usually supplied by dealers. The solution should be stirred with a wooden or glass rod while the potash is dissolving, otherwise the evolution of heat may fracture the jar. Finally, a stratum of heavy paraffin oil ($s. g.$ 1.46; 48° Baume), about $\frac{1}{4}$ in. deep, is poured upon the solution to prevent evaporation.

The cell will ordinarily require no further attention until its materials are entirely consumed, when both the zinc and oxide plates, as well as the solution, must be renewed.

60. Chemical Reactions of the Oxide of Copper Cell.—When the external circuit is closed, the oxygen of the water in the solution, uniting with the zinc, forms oxide of zinc as in other cells. This, combining with the potash in the solution, forms a soluble double salt of zincate of potash, which is decomposed as rapidly as it is formed. The hydrogen which is set free unites with the oxygen of the protoxide of copper of the negative plate, and deposits metallic copper. The reaction takes up 1 equivalent of zinc, 1 of potash, 1 of protoxide of copper, and deposits 1 equivalent of metallic copper. The wasteful local action in this cell is so small as to be practically negligible, which is an important advantage.[14] The copper is deposited in a pure form, suitable for industrial uses.

61. The Grove and Bunsen Cells.—Other voltaic combinations, formerly largely used in telegraphy but now obsolete, consist of amalgamated zinc in dilute sulphuric acid, and platinum in nitric acid known as the Grove, and carbon in bichromate of potash solution known as the Bunsen.[15]

61a. The Wasteless Battery Zinc.—The unavoidable waste of metal in the gravity cell (10) from the unconsumed part of each zinc electrode which has to be thrown aside, sometimes amounts to

FIG. 11a.
d'Infreville's Wasteless Zinc.

45 per cent. of the original weight. This loss is avoided by the "wasteless" electrode invented by G. d'Infreville, which is made up of two or more similar sections, each formed of a hub with inclined radial arms (see Fig. 11a.) The hubs of the several sections are slightly coned, and fit snugly into one another. Fig. 11b. shows an electrode of three sections after having been some time in use. When the lowermost section has been nearly consumed, a new one is added at the top, and in this way each is oxidized in turn without waste. The coned form of the hubs enables the sections to be put together in a perfectly secure manner by a light blow. With this electrode, the resistance (153) per cell is reduced to one-third its former value, while much is gained in constancy.

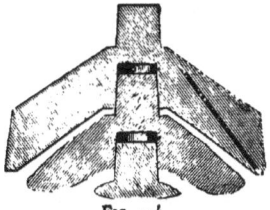

FIG. 11b.
Sectional view of Wasteless Zinc.

A brass hanger or support accompanies the zinc, which grasps it securely by an ingenious elastic friction. A plan view of this hanger

FIG. 11c.—d'Infreville's Hanger.

is shown in Fig. 11c. A connecting wire of any thickness may also be firmly clamped, as shown, between the branches of the Y-shaped extremity of the hanger, the arms of which interlock by their own elasticity so as to hold it securely.

[14] F. DeLalande and G. Chaperon in *L'Electricien*, vi. 98, 103; *Electrical Review* (London), xiii. 59, 102; xiv. 485; N. Y. *Electrical Engineer*, ix. 153.

[15] For description and directions for management of the Grove cell see *Modern Practice of the Electric Telegraph*, 4th ed, 15; and for Bunsen bichromate cell, *the same*, p 17.

CHAPTER III.

THE SOURCES OF ELECTRICITY.—(*Continued.*)

62. Magneto-Electricity.—Electricity which is evolved from a magnet, by moving coils of wire within the sphere of its influence by mechanical power, is called *magneto* or *dynamo-electricity*. The distinction between the two is purely arbitrary and nominal, and has reference only to the particular structure and organization of the machines from which they are respectively derived.

63. Magnetism.—It has been known from time immemorial that certain natural ores of iron possessed the property of attracting *iron* and *steel*, and that these metals were themselves capable, under proper conditions, of being endowed with a like property. This property, which is called *magnetism*, is also capable of being manifested, though in a less marked degree, by certain other metals, especially *cobalt* and *nickel*. Such a mass of magnetic ore is called a natural magnet or *lodestone*. A mass of iron or steel to which magnetic properties have been imparted by any known means, is called an *artificial magnet*. Soft iron is capable of retaining magnetic properties only during such time as it remains under the direct influence of the magnetizing force, and under such conditions is said to be a *temporary magnet*. Hardened iron or steel continues to retain magnetic properties after the withdrawal of the magnetizing force; and hence a mass of hardened steel, when magnetized, is called a *permanent magnet*.[1]

64. The Magnetic Needle.—A piece of hardened steel, which has been permanently magnetized, possesses marked peculiarities. When a straight bar of this kind, which is termed a *bar-magnet*, is suspended freely by its center of gravity, it always tends to place itself approximately north and south, usually in the direction of its greatest length. The imaginary line in which it thus places itself is termed the *magnetic meridian*. A small magnetic steel bar, when

---

[1] For an exposition of the modern theories of magnetism, the student is referred to the papers of D. E. Hughes, *Proc. Royal Soc.*, 1879, p. 56; J. A. Ewing, *Royal Soc.*, 1890; *Elec. World*, xvi. 241. A summary will be found in Kapp, *Electric Transmission of Energy*, 16. The celebrated lecture of Prof. A. M. Mayer, *The Earth a Great Magnet*, New Haven, 1872, presents the whole subject of magnetism in a most admirable, popular way.

suspended by a filament, as shown in Fig. 12, or upon a pivot, as shown in Fig. 13, is called a *magnetic needle*. Such a needle, in conjunction with a graduated dial, constitutes the well-known *magnetic compass*.

65. Phenomena of Magnetic Induction.—When an artificial magnet is placed in the immediate neighborhood of one or more pieces of iron, or of a quantity of iron chips or filings, these are instantly attracted. They attach themselves to the magnet, and will be found to adhere with considerable force to its surface. At the same time, a magnetic influence is exerted upon these bodies by virtue of which they themselves become magnets. The magnetism thus appearing in such bodies is said to be *induced* in them, and this process of imparting or developing magnetism is called *magnetic induction*. Thus, in Fig. 14, NS is a bar-magnet, k is an iron key which is attracted and held suspended by it, and n is an iron nail, in turn held in the same way by the key, which has itself become a magnet. The original magnetizing body suffers no loss of magnetism by this process.

FIG. 12. Suspended Magnetic Needle.

66. Polarity of the Magnet.—If a bar-magnet be rolled in a mass of filings or other small fragments of iron, these will be found to assemble in much greater

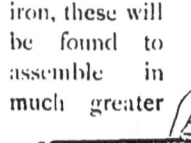

FIG. 13. Magnetic Needle on Pivot.   FIG. 14. Attraction of Magnet.

quantity near each of the ends than toward the middle of the bar, as shown in Fig. 15. This shows that the attractive force of a magnet is at its maximum at two points situated near the respective ends of the bar, and gradually diminishes toward the center, where it disappears altogether. These two points of maximum attraction are termed the *poles* of the magnet. The one which points toward the north pole of the earth when the magnet is suspended, is conventionally termed the boreal or *north pole* (71), and the opposite one the austral or *south pole*.[2]

FIG. 15. Attraction of Iron Filings by Bar-Magnet.

The intermediate point, where no attraction is manifested, is called the *neutral line* or *equator* of the magnet. Some magnets, termed *multipolar magnets*, have more than one set of poles.

The distance between the poles of a magnet is called its *magnetic length*. In most bar-magnets it is about 0.83 of the total length. In a horseshoe magnet (67) it is the shortest distance between the poles.

A magnet need not necessarily be magnetized in the direction of its greatest length; a bar may be magnetized transversely, or in fact in any direction. When a magnet is broken into detached parts, each fragment instantly becomes an independent magnet, having a north and south pole.

67. Horseshoe Magnet and Armature.—Instead of being straight, as in Fig. 14, it is more usual, as well as more convenient, for the magnetic bar to be given a form resembling the letter **U**, as in Fig. 16. This form is known as the *horseshoe magnet*. A soft iron *armature* is usually fitted to the poles of a horseshoe magnet. This is sometimes called the *keeper*, because it aids in retaining or keeping the magnetic qualities of the bar. In general terms, any mass of iron or steel subjected to the attraction of a magnet is considered to be an armature.

A magnetic attraction has been experimentally produced between a magnet and its armature as high as 1,000 lbs. per sq. in. of surface in contact.[3]

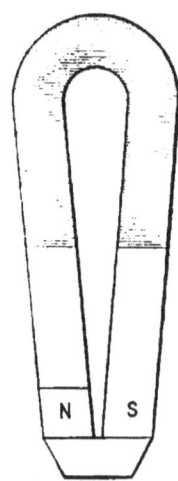

FIG. 16. Horseshoe Magnet and Armature.

68. The Magnetic Spectrum.—If a sheet of thin glass or

---

[2] The north pole of a magnetic bar or needle, by convention, is usually painted blue, and the south pole red. Sometimes they are respectively stamped with the letters N and S, and sometimes a straight line or mark serves to designate the north pole.

[3] EWING and LOW: *Phil. Trans. Royal Soc.*, 1889, A. 221; see also H. E. J. G. DU BOIS: *Phil. Mag.*, April, 1890.

card-board be laid upon a
bar-magnet, and its surface
sprinkled with iron filings
from a pepper-box, as in
Fig. 17, upon tapping the
sheet with a pencil or similar object, a remarkable
phenomenon will occur.
The particles of iron will
arrange themselves symmetrically in curiously
curved lines as shown in

FIG. 17. Method of producing Magnetic Spectrum.

Fig. 18, which is taken from a photograph. This is called the *magnetic spectrum*.

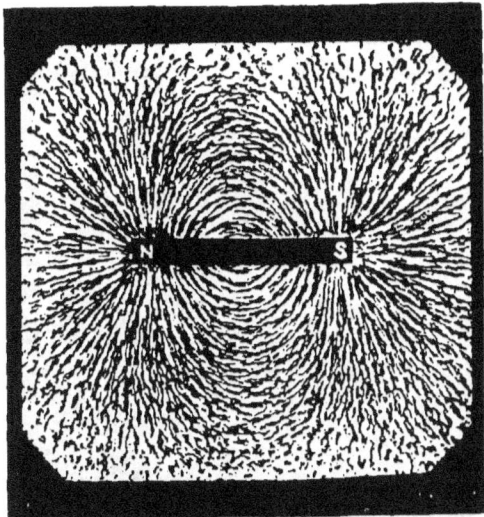

FIG. 18. Spectrum of Bar-Magnet.

69. **The Magnetic Field.**—The sphere of attraction which surrounds a magnet is termed the *magnetic field*, and is filled with what were happily termed by Faraday, *lines of magnetic force.* These exist unseen in every magnetic field, but their presence and direction may be made evident by the expedient which has just been described. Magnetic force in itself is absolutely inappreciable by any of our senses. We only know of its existence by its effects upon matter.

Since the peculiarities of the magnetic field are due to the presence of a force, the properties of such a field may be made known by determining the strength and the direction of this force, or, as it is usually expressed, the *intensity of the field*, and the *direction of the lines of force.*[4]

[4] Force is any action which can be expressed simply by weight, and is distinguished by a great variety of terms, such as attraction, repulsion, gravity, pressure, tension,

28   *Sources of Electricity.*

70. **Lines of Magnetic Force.**—The invisible lines of magnetic force radiate in every direction from each pole of the magnet. They may be regarded as an inseparable part of it, which accompany it wherever it goes. Perhaps their true nature may be more clearly conceived by assuming them to set out from one pole, say the north pole, and after curving for a greater or less distance through space, to return again to the south pole, as indicated by the arrows in Fig. 19. A view of the spectrum of the magnetic field at one pole of a bar-magnet, as seen end-on, exhibits merely radial lines, as in Fig. 20.

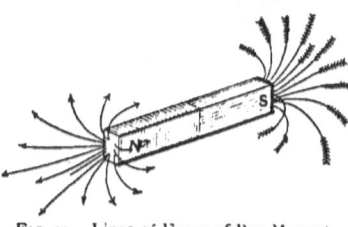

FIG. 19. Lines of Force of Bar-Magnet.

71. If a small bar-magnet or magnetic needle be suspended at any point within the field of a larger magnet, it will invariably tend to place itself parallel to the line of force which passes through both its poles, as shown in Fig. 21. This explains why the needle of the magnetic compass always points to the north. The earth itself is a great magnet, and is surrounded by a field filled with invisible lines of force which we term *magnetic meridians*. These lines determine the position of the suspended magnetic needle. Thus by exploring with such a needle, the direction of the lines of force in any magnetic field may be discovered (94).

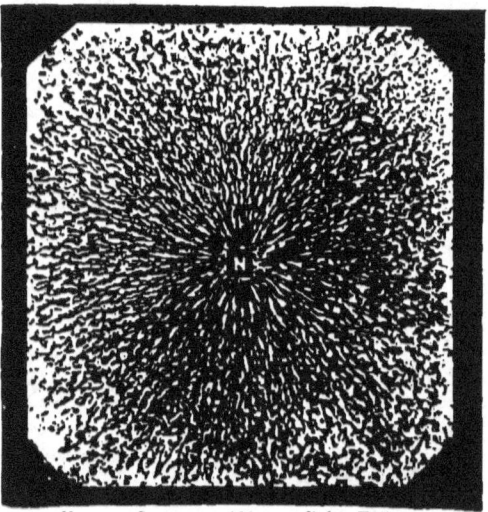

FIG. 20. Spectrum of Magnet Pole—End-on.

72. **Attraction and Repulsion.**—The respective north poles

compression, cohesion, adhesion, resistance, inertia, strain, stress, strength, thrust, load, squeeze, pull, push, etc., all of which can be measured or expressed by *weight*, without regard to motion, time, power or work.—J. W. NYSTROM: *Elements of Mechanics*, p. 59.

Current Produced by a Magnetic Field. 29

of any two magnets *repel* each other, and so do the south poles ; but, on the contrary, the north and the south pole of the same or differ-

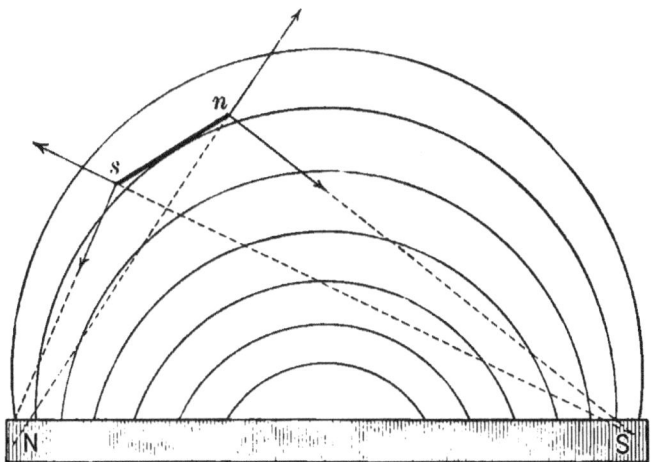

FIG. 21. Position of Magnetic Needle in Field.—J. A. Fleming.

ent magnets mutually *attract* each other. Hence it follows that the north pole of any magnet must have the same polarity as the south pole of the earth, and in strictness ought to be termed the south-pole rather than the north (66). It is more properly termed the *north-seeking* pole.

73. **Electric Current Produced by a Magnetic Field.**—If a conducting wire in the form of a closed loop or endless ring be moved within a magnetic field, in any direction whatsoever which alters the number of lines of force passing through it, a *current of electricity* will appear in the wire. The same thing will occur if the wire be stationary and the field be moved; or if the wire be stationary and the intensity or strength of the field be increased or diminished, either between zero and maximum, or to a lesser extent; or if the wire be moved from one part of the field to another part of different intensity.[5]

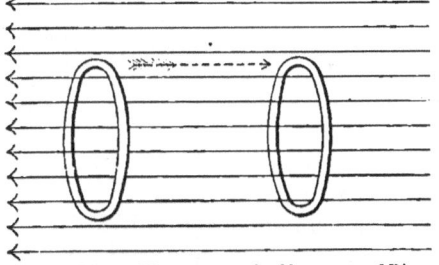

FIG. 22. Lines of Force not cut by Movement of Ring.

[5] FARADAY's own account of this capital discovery of *magneto-electricity*—the results of which are likely to ultimately become of greater importance than any other ever

74. Thus in Fig. 22,[6] let the parallel arrows be assumed to represent lines of force in a uniform magnetic field. If the closed ring of wire be moved parallel to those lines, as indicated by the dotted arrow, no electric current will appear in the ring. Or if the ring and the lines of force, either or both, be moved in a transverse direction with reference to each other, without altering the total number of lines enclosed, as shown in Fig. 23, no current will be generated in the ring. Fig. 24, on the other hand, represents a field which is not uniform, being stronger or more intense, or in other words, having a greater number of lines of force, in some parts than in others. Moreover, as shown by the arrow-heads, these lines run in opposite directions in different parts of the field. If, now, the ring be moved from a place where the lines of force are more numerous to a place where they are less numerous, as from position 1 to position 2 in Fig. 24, a current will be generated; and if this motion be continued, as in position 3, to a place where the lines run in an opposite direction, the effect will be similar in kind, but will be even greater in amount. So, also, if the ring be moved in a uniform field in such a manner that either the number or the direction, or both, of the lines of force cut by it are varied, a current will be produced. This happens if the ring be turned round an axis at right angles to the direction of the lines of force, as shown in Fig. 25.

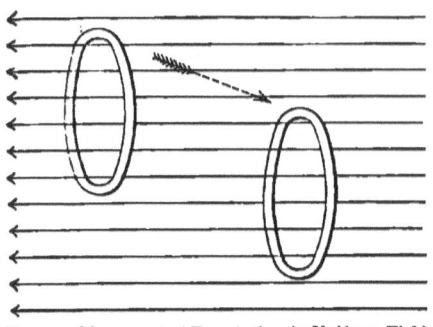
FIG. 23. Movement of Translation in Uniform Field.

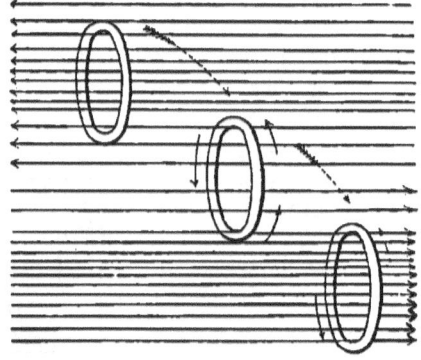
FIG. 24. Movement of Ring in Field of Varying Intensity.

---

achieved by man, with the possible exception of the discovery of the expansive power of steam—is given in his *Experimental Researches*, i. 7. See *N. Y. Elect. Eng.*, xiii. 27.

[6] SILVANUS P. THOMPSON: *Dynamo-Electric Machinery* (2nd edition), 12.

This last described organization is very common in machines for producing electricity from magnetism.[7]

75. **Transformation of Mechanical Power into Electricity and Heat.**—In thus moving a closed circuit or loop of wire through a magnetic field so as to cut across the lines of force, a certain physical resistance is encountered, and a corresponding mechanical force must be applied to overcome it and effect the motion. The *equivalent* of mechanical energy thus consumed reappears as electricity in the closed ring, except that a certain portion, which is transformed into heat, as will be hereafter more fully explained (87).

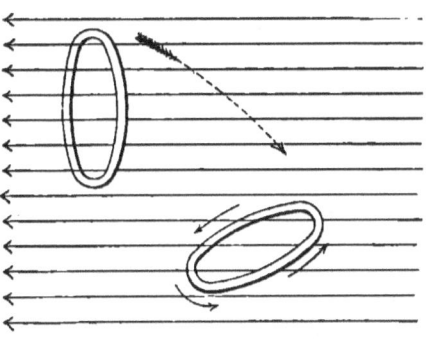

FIG. 25. Circular Movement of Ring in Uniform Field.

76. **Direction of the Induced Current.**—It has been stated (31), that what we call the *direction* of a voltaic current is conventionally assumed to be from the positive pole of the cell through the conducting wire to the negative pole; and it will be obvious that if the respective poles were interchanged, the current would traverse the wire in the opposite direction. The direction of the current produced in a conductor by moving it with reference to the lines in a field of force, called the *magneto-electric current*, depends upon the direction in which the relative motion takes place. The law may be stated as follows:

A *decrease* in the number of lines of force which pass through or are cut by a closed circuit, produces a current round that circuit in the *positive* direction; while an *increase* in the number of lines of force which pass through or are cut by such circuit, produces a current around such circuit in the *negative* direction.[8]

The positive direction of the lines of magnetic force which pass through the loop of the circuit, is invariably associated with a positive direction of the current flowing round the conducting circuit,

---

[7] This and the four following paragraphs explanatory of the mutual reactions of the magnet, the magnetic field and the conductor, are abridged from a portion of chapters ii. and iii. of SILVANUS P. THOMPSON's admirable work on *Dynamo-Electric Machinery*.

[8] SILVANUS P. THOMPSON: *Elementary Lessons in Electricity and Magnetism*, 360.

just as the forward thrust is with the right-handed rotation in the operation of driving an ordinary right-handed screw. This will appear from an examination of the direction of the current in the ring as shown by the arrows in Fig. 25.

77. Mutual Reactions of a Current and a Magnet.— The phenomena which have been described, like most physical phenomena, are *reversible;* that is to say, a magnetic field may also be created by the passage of an electric current through a wire conductor, and, moreover, a mass of iron or steel situated in such a field will become magnetic. This effect, which is called *electro-magnetism* (86, *d*), lies at the foundation of electric telegraphy.

78. Summary of Magneto-Electric Phenomena.—It appears, therefore, from the facts which have thus far been stated, that:

(i.) An electric current is set up in a conductor, either by moving a magnet near such a conductor, or by moving the conductor near a magnet.

(ii.) The establishment and maintenance of a continuous electric current in a conductor requires a continuous expenditure of energy, or, in other words, consumption of power, in order to produce the necessary motion.

(iii.) To induce currents in a conductor, there must be relative motion between conductor and magnet, of such kind as to alter in some manner the number of lines of force cut transversely by the conductor.

(iv.) Other things being equal, the more powerful the magnetic pole or magnetic field, the more powerful will be the electric current generated.

(v.) The more rapid the relative motion of the two elements the more powerful will be the current.

(vi.) The direction of the induced current depends upon the direction of the motion of the wire with reference to the direction of the lines of force in the field.

79. The Dynamo-Electric Machine.—A dynamo-electric machine, briefly termed a *dynamo*, in the general and most proper sense of the term, embraces every machine capable of converting the energy of mechanical motion into the energy of an electric current, and it is in this sense that the term dynamo will hereafter be used in this treatise.

80. The Theoretical Dynamo.—The simplest conceivable dynamo is illustrated in Fig. 26. It consists of a single rectangular loop of wire, rotating in a uniform magnetic field maintained between the north and south poles of a large horseshoe magnet. If the loop be first placed in a vertical plane, as in the figure, the number of lines of force passing through it will be a maximum, but as it is turned by the crank into a horizontal position, the number of intersecting lines will obviously diminish to zero. On continuing the rotation beyond this point, the lines begin again to thread through the loop from the opposite side, so that there will be a negative or reverse maximum when the loop has been turned through an angle of 180°, or half-way round. During the first half of the revolution, therefore, a current will be induced in the loop in one direction, the strength of which will increase gradually from zero to maximum and then diminish again to zero. Upon passing the 180° position, there will begin an induction in the reverse sense, and a similar effect will

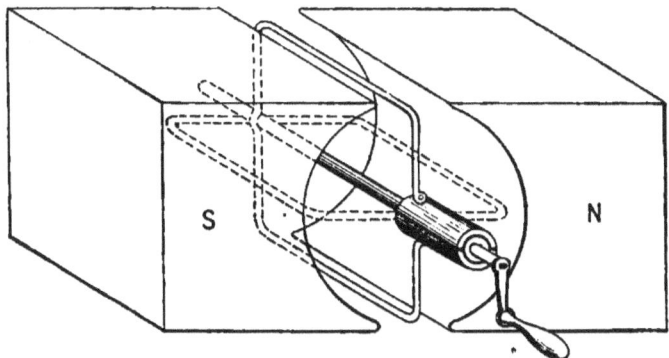

FIG. 26. Theoretical Dynamo-Electric Machine.

again take place, resulting in the induction of a current in the loop in the opposite direction, the operation being completed when the loop has been carried through one complete revolution.

81. Further considerations, having reference to the construction and mode of operation of the actual dynamo, may profitably be postponed until the student has become familiar with the fundamental laws which govern the flow and distribution of electric currents, as the reaction of these currents upon the machine which produces them, though important, are somewhat complex, and in the absence of such knowledge will be found difficult of comprehension (299).

82. **Frictional Electricity.**—Frictional electricity, which is one form of what is termed *static electricity*, finds no practical application in telegraphy. Nevertheless, the phenomena of static electricity, under certain conditions, manifest themselves in such a way as to

interfere with the transmission and reception of telegraphic signals, and hence it will become necessary, in connection with that subject, to give the matter further consideration. (See Chapter VIII., § 314.)

83. Thermo-Electricity.—This name has been given to electricity derived from the direct conversion of heat-energy. Its application to telegraphy has thus far been purely tentative, and does not require consideration here.[9]

---

[9] For an account of thermo-electric apparatus, and its experimental application to telegraphy, consult LOCKWOOD: *Electricity, Magnetism, and Electric Telegraphy*, chap. iii. p. 36.

CHAPTER IV.

THEORY OF QUANTITATIVE ELECTRICAL MEASUREMENT.

85. The Electric Current.—It has been stated (28) that a conducting wire uniting opposite poles of what, for convenience, we call a generator of electricity, whether this be a voltaic cell (30) or a magneto-electric apparatus (80), is endowed with certain peculiar properties, by reason of which we conventionally assume an electric current to flow from the positive to the negative pole of such generator.

86. Manifestations of the Current.—The existence of this so-called electric current in the conjunctive wire is manifested in several different ways, the most important of which are as follows:

FIG. 27. Iron Filings held to Wire by Magnetism.

(*a*) If the wire be dipped in iron filings, a mass of these will cluster around it and apparently adhere to it, appearing as in Fig. 27.

(*b*) If the wire be placed in the immediate vicinity of a freely suspended magnetic needle (64), the latter will immediately tend to set itself at right-angles thereto, as indicated in Fig. 28. Moreover, the *direction* in which the needle moves will indicate the direction of the current (31).

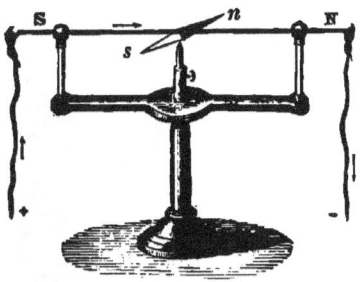

(*c*) If the wire be placed parallel to another wire, or to another portion of the same wire which is also conveying an electric current, repulsion or attraction will be manifested between the two wires according as the two currents flow in the same or in opposite directions.

FIG. 28. Deflection of Needle by Current.

(*d*) If the wire be wound spirally around a rod of soft iron, as in Fig. 29, the iron will become a magnet, and will continue to be magnetic so long as it remains under the influence of the current, but upon the removal of the wire or the cessation of the current, nearly every trace of magnetism will disappear from the iron.

FIG. 29. Magnetization of Soft Iron by Current.

35

(*e*) If the wire be severed and its ends immersed in water, the water will be decomposed, the oxygen appearing at one terminal and the hydrogen at the other, as shown in Fig. 30. This action is termed *electrolysis*.

(*f*) If the severed ends of the wire be united by a very thin wire of platinum, 3 in. or 4 in. long, and this be placed in a vessel of alcohol, a thermometer will show the liquid to become heated by the action of the current upon the wire (see Fig. 31).

(*g*) If the ends of the severed wire be placed side by side upon the tongue, a peculiar taste will be experienced; and if the current be strong enough, it may be felt by the fingers. This sensation is termed an *electric shock*.

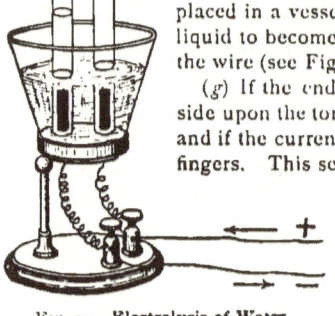

FIG. 30. Electrolysis of Water.

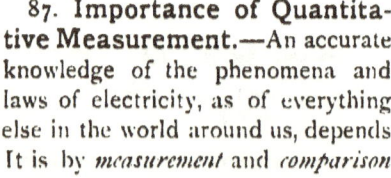

87. Importance of Quantitative Measurement.—An accurate knowledge of the phenomena and laws of electricity, as of everything else in the world around us, depends primarily upon measurement.[1] It is by *measurement* and *comparison* alone that we are able to understand electrical phenomena.

88. Fundamental Units of Mass, Space, and Time.—The principle of the

[1] In physical science, a first essential step in the direction of learning any subject, is to find *principles of numerical reckoning*, and *methods of practically measuring some quality connected with it.* I often say that when you can measure what you are speaking about and express it in numbers, you *know* something about it; but when you cannot measure it, when you cannot express it in numbers, your knowledge is of a meagre and unsatisfactory kind; it may be the beginning of knowledge, but you have scarcely in your thoughts advanced to the state of *science*, whatever the matter may be.—SIR WILLIAM THOMSON: *Popular Lectures and Addresses.* p. 73.

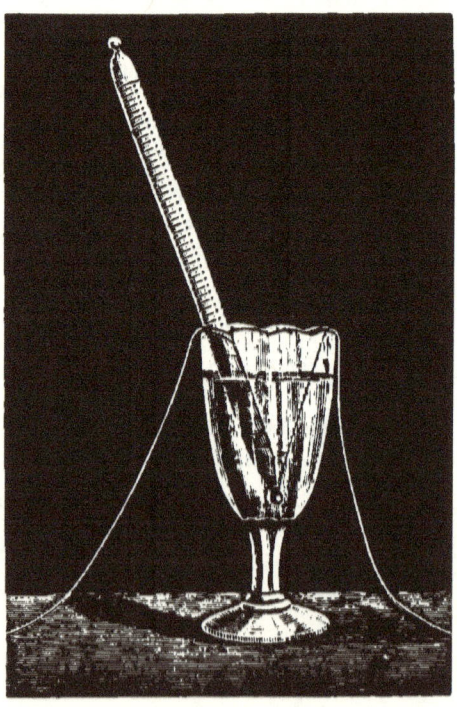

FIG. 31. Development of Heat by Electric Current.

absolute measurement of electricity and magnetism is, as Thomson remarks, "merely an extension of the astronomer's method of reckoning mass in terms of what we may call the universal gravitation unit of matter, and of the reckoning of force adopted by astronomers, in common with all workers in mathematical dynamics, according to which *the unit of force is that force which, acting on a unit of mass for a unit of time, generates a velocity equal to the unit of velocity.*[2]

89. **Illustration of the Absolute System of Measurement.**—As a concrete example, suppose we take a pound weight as our *unit of mass*, and allow it to drop through space for a period of one second, our *unit of time*. This mass will always fall through the same space during the unit of time, and at the end of that time will be capable of striking with a certain determinate force, which is obviously measurable by an equivalent weight, and which therefore becomes our *unit of force*, while the distance through which the mass falls in one second becomes our *unit of space*.[3] Such a system of measurement being wholly independent of the physical properties of any arbitrary material, is properly called an *absolute* system.

90. **Derivation of Electrical and Magnetic Units.**—The actual units used in the measurement of electricity and magnetism, are founded upon the French or metric system of weights and measures, which has been commercially adopted by all the civilized countries of the world except Great Britain and the United States, and is in extensive use in the last named countries among scientific men. In electro-magnetic measurement, therefore, the *centimetre* has been adopted as the unit of space, the *gram* the unit of mass or weight, and the mean solar *second* the unit of time. This is briefly denominated the *c. g. s.* (centimetre-gram-second) system.[1]

91. **The C. G. S. Units of Force and Work.**—The act of moving a weight of 1 gram through a space of 1 centimetre, during the time of 1 second, requires a perfectly definite and measurable

---

[2] The units of *space*, *mass*, and *time*, have been selected by common consent to serve as *fundamental units*. Other units, for practical use, determined from these, such for example as the unit of *force*, are termed *derived units*.

[3] In making this general statement, the effect of the resistance offered by the air has been neglected (this being, of course, greater for a less dense than for a more dense body), as has also the fact that a given mass which weighs a pound, for example at Washington, D. C., will weigh more than a pound at the north or south pole of the earth, and less than a pound at Panama, or at the equator. This is due to the fact that the earth is not a perfect sphere.

[4] The centimetre is somewhat less than half an inch English measure; 1 foot is very nearly 30.5 centimetres; 1 cubic centimetre of water weighs 1 gram; 1 oz. is very nearly 28 grams; 1 lb. is 454 grams; the 5-cent nickel of the 1873 U. S. coinage weighs exactly 5 grams and has a diameter of 2 centimetres; the silver dime weighs 2.5 grams.

amount of force, which is termed a *dyne.* The dyne, therefore, is the *unit of force* in the *c. g. s.* system, and is defined as *the force, which acting upon a gram for a second, generates a velocity of a centimetre per second.*[5] Any force may be stated to be equal to so many dynes. A *megadyne* is equal to 1,000,000 dynes (123, *note*).

The *erg* is the unit of *work* in the *c. g. s.* system, and is *the work done by a force of* 1 *dyne acting through a distance of* 1 *centimetre, irrespective of the time occupied.*[6]

Now if a force of 1 dyne be applied to move a closed conducting ring a distance of 1 centimetre through a uniform magnetic field (as, for example, the magnetic field of the earth), in the manner explained in (73), *work* is done; an electrical force is set up in that conductor which would be the exact *electrical equivalent* of the mechanical force of 1 dyne, were it not that some part of the original force is unavoidably transformed into heat during the operation. Neglecting the value of the heat-loss, this quantity of electricity is capable of doing mechanical work equal to 1 *erg*, as, for instance, by forcibly deflecting a magnetic needle (86, *b*) or by attracting the armature of a magnet (86, *d*).

92. **The Conservation of Force.**—This is one illustration of the great principle of the indestructibility or, as it is commonly called, the *conservation of force*,[7] which is so important, that it has been justly remarked that the whole of natural philosophy is merely a commentary upon it.[8]

It follows from this principle, that whenever a signal is produced at any point by electrical action, a physical effect must be made to take place, and this necessarily involves the expenditure of some form of force at some other point. It may be the force of chemical affinity in the voltaic cell; or it may be the force of steam or of falling water, or of human muscles exerted upon a dynamo-electric

[5] EVERETT: *Units and Physical Constants*, pp. 22, 167.
[6] EVERETT: *Units and Physical Constants*, p. 167.
[7] DANIELL: *Principles of Physics*, 7.
[8] This doctrine teaches that the total amount of force in the universe is unalterable, and that it can neither be created nor destroyed. Force, however, may appear in a variety of different forms, and is capable of being readily changed from one form to another, but every such mutation is nevertheless rigidly subject to quantitative laws. *A given amount of one form of force produces a definite quantity of one or more other forms of force and no more.* Hence this law is sometimes called the *equivalence of forces*. This important and interesting subject is well worthy of further study, and among special works relating to its various aspects, the author ventures to specially commend the following:—TYNDALL: *Heat as a Mode of Motion;* YOUMANS: *The Correlation and Conservation of Forces;* a collection of papers by Grove, Helmholtz, Mayer, Faraday, and others; STEWART: *On the Conservation of Energy;* and SPRAGUE: *Electricity, its Theory, Sources, and Applications.*

Electric Field of Force. 39

machine, but in every case, the force expended must be equal to that which is utilized, plus that which is transformed into heat in the course of the operation. Telegraphic signals are usually produced by means of the attraction of an armature by an electro-magnet. *The initiation and maintenance of this attraction involves the consumption in the battery of a perfectly definite and well ascertained quantity of material, which can never be less than the full equivalent of the mechanical work done*, but must be somewhat more, and may possibly be very much more, for by unskillful arrangements an undue proportion of the original force may be turned into heat and rendered unavailable for the purpose in hand (154).

93. **Electric Field of Force.**—If we take a magnetic needle, which, as we have seen, tends to remain in the magnetic meridian, and place parallel to it a wire, traversed, as shown in Fig. 28, by a current in the direction of the arrows, the needle will seek to place itself at right-angles to the wire (86, *b*), but being under the influence of two antagonistic forces, it will come finally to rest in an intermediate position. This experiment shows that a conductor, when con-

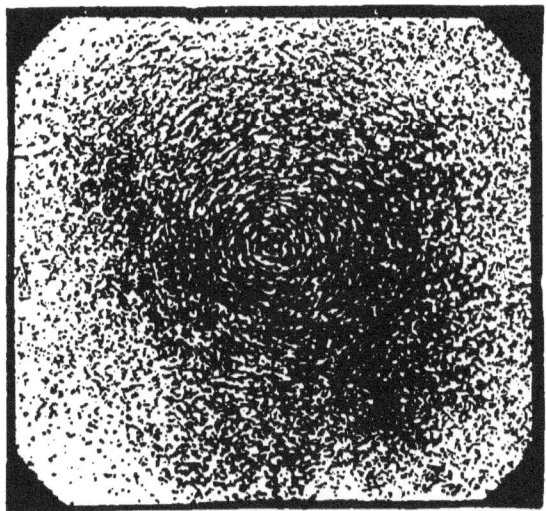

FIG. 32. Spectrum of Field Surrounding Conductor.

veying a current, is surrounded by a field-of-force of the same character as that which we have found to surround the magnet (69). To determine the position and direction of the lines of force in this field, we may adopt the same expedient as in the case of the magnet.

Pass the conducting wire at right-angles through a piece of glass or card-board. If iron-filings be dusted into the field, they will arrange themselves in concentric circles (Fig. 32), showing that the lines of force *encircle* the wire, instead of *radiating* outwardly from it as they did in the case of the magnet. It is these lines of force which act upon the needle and tend to set it at right-angles to the wire, for when any concentric line passes through both poles of the needle, the latter must set itself at right-angles to the radii of the circle (71).

94. Relation of Current Force to Mechanical Force.— The particular angular position of a needle under the influence of a current must necessarily depend upon the ratio between the strength of the magnetic field-of-force due to the *current*, and that of the field-of-force due to the *magnetism of the earth*, which may be regarded as sensibly constant in any particular locality.

That portion of the force of the earth's magnetism which acts to hold a horizontal needle in the meridian is called its *horizontal component* (H). Its value varies from a maximum at the magnetic equator to nothing at the magnetic poles. Its locality of greatest intensity is in lat. 0° and long. 101° W., where it is equal to 0.3733 dynes. Following are some determinations of its value by observers of the U. S. Coast and Geodetic Survey:

Washington, D. C.	0.2026	dynes.
New York	0.1872	"
Eastport, Me.	0.1573	"
Key West, Fla.	0.3055	"
Cincinnati, O.	0.2111	"
San Francisco, Cal.	0.2533	"
St. Paul Island, Alaska	0.2008	"
Toronto, Ont.	0.1654	"
City of Mexico	0.3429	"

(*Rep't U. S. C. & G. S.*, 1885; App. No. 6.) This report gives the value of the horizontal intensity found in over 1,500 observations in various parts of North America, reduced to the epoch of 1885.

The value of the horizontal intensity is subject in most places to a slow annual variation. Along a line drawn from British Columbia to Florida, the intensity is constant; east of this line it shows an annual increase, and west of the line an annual decrease. [*Rep. U. S. C. & G. S.*, 1885, p. 271, and charts.][*]

As the strength of the field, due to the current, is always strictly proportionate to the capacity of the current to produce other physical effects, we have a means, not only of comparing the forces of differ-

---

[*] The refined methods of determination used are fully described by C. A. SCHOTT in App. 8 of *the same* for 1881. For an elementary explanation of these methods see TROWBRIDGE: *New Physics*, p. 142. See also A. GRAY: *Absolute Measurements in Electricity and Magnetism*, p. 5; F. E. NIPHER: *Theory of Magnetic Measurements*, p. 46.

ent currents, but, by comparison with the earth's magnetism, of determining the *actual dynamic value* of any current, in terms of the fundamental units of *space, mass,* and *time* (91).

A unit magnetic pole weighing 1 gram, and free to move in a horizontal plane, under the action of the earth's horizontal force, would require, at the end of 1 second, a velocity equal to 202.6 centimetres per second, if the experiment were made in Washington, D. C. (See p. 40.)

95. The Galvanoscope and the Galvanometer.—A magnetic needle provided with a conductor through which a current may be passed in order to deflect it from the meridian (86, *b*), is called a *galvanoscope* or *detector*. When to these is added a graduated scale or dial, the instrument becomes a *galvanometer*. In order that the angle of deflection of a needle under the influence of any current shall bear a definite ratio to the value of such current, certain precautions in its mechanical construction are necessary to be observed. It is essential that the field produced by the current should, like that of the earth, be so large in comparison with the needle, that the motion of the latter within it, when deflected, shall not appreciably change its relation to the entire field.

96. Tangent Galvanometer.—In this instrument the foregoing condition is fulfilled. In its most simple form, the tangent

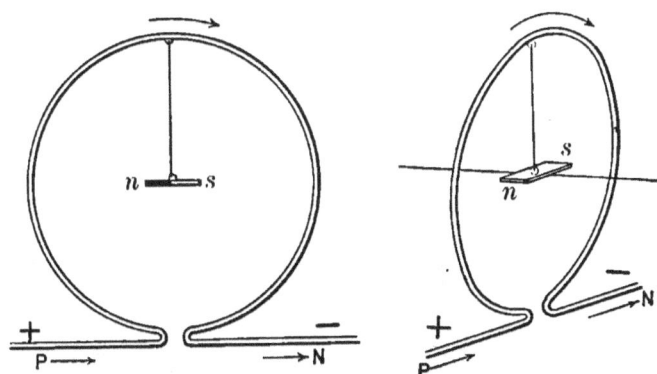

Fig. 33. Needle in Circular Loop.

galvanometer consists of a single circular turn or loop formed in the conducting wire; in the center of the loop is suspended the needle, which in length should not exceed $\frac{1}{10}$ the diameter of the loop. Such an organization is shown in Fig. 33, in which *n s* is the suspended needle, and P N the looped conductor which surrounds it.

Quantitative Electrical Measurement.

The principle of action may be understood by reference to Fig. 34. Let the magnetic needle $n\,s$ be suspended in the earth's magnetic meridian N S. If now the conducting wire or loop be placed in the plane of N S, and we suppose this wire to be traversed by a current capable of producing a magnetic field precisely equal in strength to that of the earth, the needle will swing toward a position represented by the line A B, at right angles (or 90°) to the one originally occupied. But the two antagonistic forces being equal, the needle will come to rest in a position half-way between N and B, called the *resultant*. This coincides with the line A1, which forms an angle of 45° with the zero or 0° line N S. Now if we double the strength of the current, and consequently that of the field produced by it, it will partially overpower the earth's field, and the needle will assume the position corresponding to the line A2, which is an angle of $63\frac{1}{2}°$ nearly. If we again double the strength of the current, we shall increase the deflection to 76°, represented by the line A4. In geometrical language, the line N4 is termed a *tangent* to the *arc* or *quadrant* N B of the circle. The circle being divided into 360°, the tangent, as found in a computed table of natural tangents (see

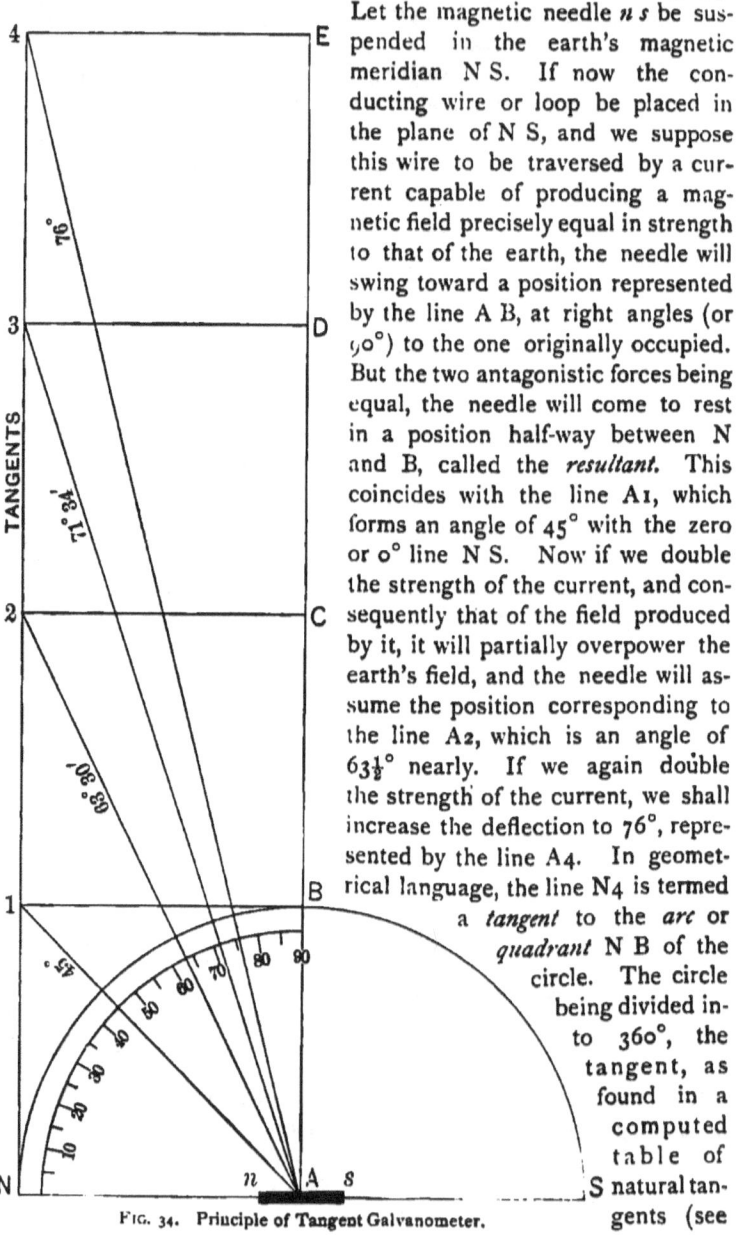

FIG. 34. Principle of Tangent Galvanometer.

page 55), will always be proportionate to the strength of the current by which the corresponding deflection was caused. The actual construction of this useful instrument will be described in detail elsewhere (102).

97. **Character of Electrical Measurements.**—The qualities of an electric current by virtue of which, as we have seen, it is enabled to exert force, to produce physical effects, or, in technical language, to do *work*, are three in number, viz: (1) *quantity*, variously called volume, or strength of current; (2) *potential*, variously called pressure, tension, and intensity,[10] and (3) *duration*, or time occupied in doing the work. The value of the first two of these properties, in the case of any particular current, is dependent not only upon the circumstances of its origin, but upon the special characteristics of the conductors which the current is compelled to traverse. Hence there are two distinct classes of electrical measurements: (1) those which are applied directly to electricity itself, either in a static or in a dynamic condition, that is to say, as a stationary charge or as a flowing current, and (2) those which are applied to the conductor which a current is or may be compelled to traverse.

98. **Characteristics Capable of Measurement.**—Electricity itself, whether in a static or a dynamic condition, has but three properties susceptible of quantitative measurement, viz: (1) *quantity*, (2) *potential*, and (3) *duration*. Electric conductors have four qualities which may affect the value of the currents which traverse them, and which, in like manner, are susceptible of measurement, viz: (4) *magnitude* (length, breadth, and thickness), (5) *weight*, or mass, (6) *temperature*, and (7) *conductivity*, or the reciprocal of this, called *resistance*. All these considerations must be taken into account in performing any electrical measurement.

99. **Apparatus for Measurement.**—A complete apparatus for executing electrical measurements must therefore comprise:

   (*a*) A meter for *quantity* of current.
   (*b*) A meter for *potential* of current.
   (*c*) A chronometer for *time*.
   (*d*) A scale for *linear* measure.
   (*e*) A scale or balance for *weight*.
   (*f*) A thermometer for *temperature*.
   (*g*) A standard of electrical *conductivity* or resistance.

---

[10] The French scientific writers have always been accustomed to use the term *intensité* in the sense in which we use the term *quantity* or *volume*. The frequent translation of this word by the English term "intensity," which has in fact a wholly different signification, has been the cause of no little confusion in electrical literature.

Not all these instruments are required in making every measurement, for one or more of the unknown conditions may be arbitrarily assumed, or they may be determined from others which are known, as we shall hereafter frequently have occasion to note.

100. **The Ammeter, Voltameter, and Calorimeter.**—The instrument for measuring quantity of current need not necessarily be a galvanometer, although in telegraphic work this is used practically to the exclusion of everything else. Currents may also be measured by the attractive force of an electro-magnet (83, *d*), as in the instrument called the *ammeter*, or by ascertaining the volume of gas evolved in a unit of time (83, *e*), in which case the instrument is called a *voltameter*, or by measuring the heat developed in a unit of time (83, *f*), in which case it is called a *calorimeter*. All these instruments find frequent use in general electrical investigations, but are less convenient than the galvanometer for measurements in connection with telegraphy.

CHAPTER V.

THE LAWS AND CONDITIONS OF ELECTRICAL ACTION.

101. Apparatus Required by the Student.—The student who desires to obtain a thorough knowledge, not only of the art of telegraphy, but of the principles of physics and chemistry upon which that art is based, is earnestly advised to supply himself, in the first instance, with his own apparatus. The comparatively small cost of the necessary outfit will be many times repaid in the value of the clear and definite experimental knowledge which this means alone will enable him to acquire. It is for many reasons advisable that two students should work together, as experience has shown that the study is rendered far more interesting, and that much more rapid and intelligent progress may be made in this way. The following list of apparatus and supplies will serve the requirements of two students, the approximate cost of each item being given:

4 cells "crowfoot" battery complete, with 4 copper connectors (9)	$0.75	$3.00
2 keys (258)	1.75	3.50
2 5-ohm sounders (261)	2.25	4.50
2 lbs. No. 18 "burglar alarm" insulated copper wire	.35	.70
2 feet ⅜-in. rubber tubing for gravity cell (39)		.30
1 2-in. glass funnel for gravity cell (39)		.10
1 hydrometer (18)		.50
1 battery brush (40)		.25
1 6-in. permanent magnet (67)		.60
5 lbs. sulphate copper	.10	.50
1 box No. 2 office-wire staples		.10
Total for two students		$14.05

102. Construction of the Tangent Galvanometer.—A tangent galvanometer and rheostat are almost absolutely necessary in the experimental investigation of electrical action, and hence it is to be regretted that a sufficiently cheap but good apparatus of this kind has never been made available for the use of students and amateur electricians. It is nevertheless quite possible for an ingenious person, somewhat accustomed to mechanical tools and processes, to

46   Laws and Conditions of Electrical Action.

construct, at a trifling cost, apparatus sufficiently accurate for all practical purposes.

The following directions for making a tangent galvanometer are taken in part from S. R. BOTTONE's *Electrical Instrument Making for Amateurs*.

103. First a ring of hard wood C is accurately turned in a lathe, of the form and dimensions shown in Figs. 35 and 36. The channel

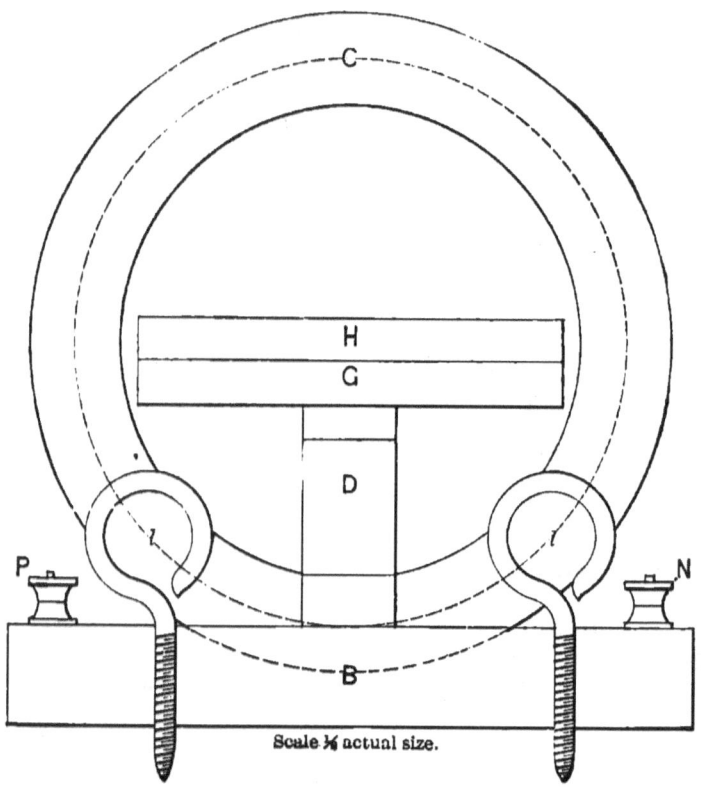

FIG. 35. Elevation of Tangent Galvanometer.

in the edge of the ring must be of such depth that the bottom of it will be exactly 6 in. in diameter and ½ in. in breadth. Such a channel will hold 38 turns (in 2 layers) of No. 22 "American gauge" double cotton-covered copper wire. The coil will require 62 feet of wire, weighing about 2 oz., and costing about 25 cents. The wire must be wound carefully and accurately in the groove or channel (this may be best done in a lathe), and the ends brought

Construction of the Tangent Galvanometer. 47

out through small holes in opposite sides of the ring and through the base, as shown in Fig. 36. The ring and the wire, being first well dried, should be thoroughly coated with shellac varnish.

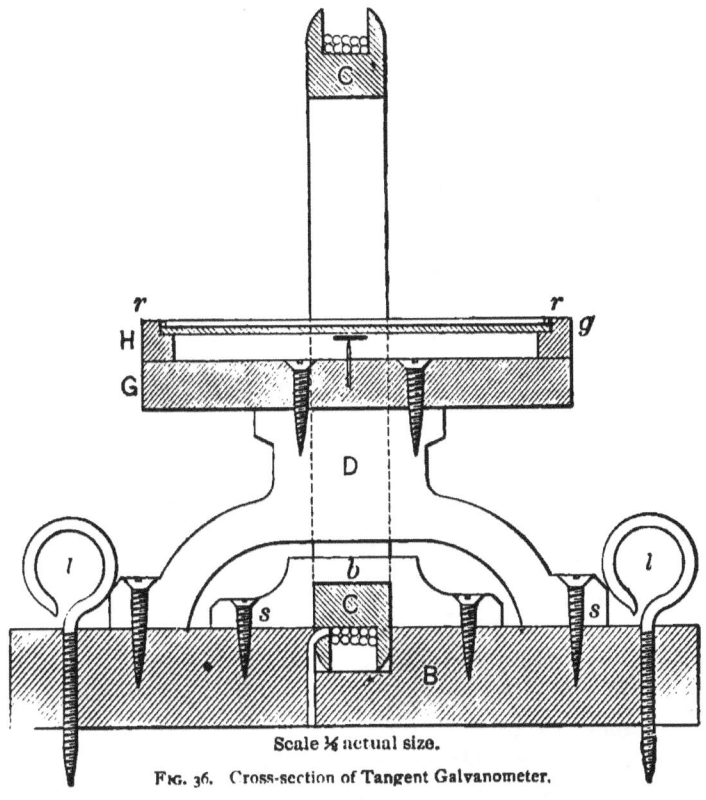

FIG. 36. Cross-section of Tangent Galvanometer.

The base B may be of hard wood, preferably turned, 7 in. in diameter and 1 in. thick. It is supported upon 3 equidistant leveling screws $l\,l$, which may be common screw-eyes (preferably of brass, though iron will answer), with the sharp tips filed off.

The ring C is fixed firmly in a vertical position upon the base B, taking care that it is placed accurately at right angles therewith. A recess may be cut in the base, the ring being let into it, and secured by a clamping piece b of wood, made fast to the base B by two brass screws s.

The magnetic needle. N S, Fig. 37, may be of a piece of watch-spring 1 in. long and $\frac{1}{8}$ in. wide. Soften this by holding it in the flame of a spirit-lamp until of a dull red color, and then allow it to

48 *Laws and Conditions of Electrical Action.*

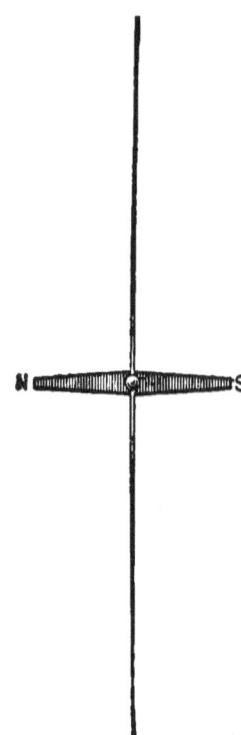

Fig. 37. Needle and Index of Galvanometer (Full Size).

cool slowly. Drill a hole exactly in the center about $\frac{1}{18}$ in. in diameter, and file the ends into the tapering form shown. Straighten the needle with a hammer, and unless its center of gravity corresponds with the center of the hole, correct the error by filing the heavy part away. Next harden the needle by reheating it in the lamp-flame until nearly red-hot, and dropping it into cold water. It must then be magnetized, by rubbing each half separately from the center to the end, with a permanent magnet (67), being very careful to rub one end with one pole of the magnet, and the other end with the opposite pole.

A jeweled center is then fitted to the hole in the needle,[1] and secured with glue, or better, with white or red lead used as a cement. It is not difficult for the amateur to make a center out of glass, in case a jeweled one cannot be procured, by holding a piece of $\frac{1}{4}$ in. glass tube in the spirit-lamp flame, and pulling it apart lengthwise as soon as it softens. In Fig. 38 this operation is illustrated at C. E is one of the two pieces, which will be drawn out as shown, into a thread-like extremity. Fuse this thin end in the flame and a little globule will be formed, closing the bore of the tube as at F. When cold the tube can be broken off by cutting a scratch at the desired point with a triangular file. This leaves a center, G, adapted to support the needle freely upon a pivot as shown in Fig. 13, page 25.

Having fixed the center in the hole in the needle as directed, poise the whole upon the point of

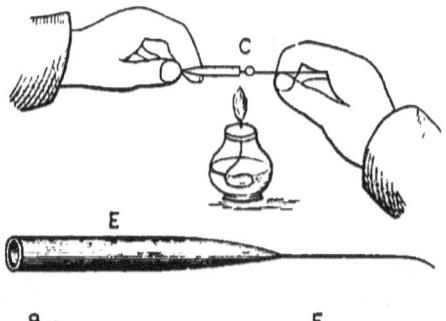

Fig. 38. Method of making Glass Center.

[1] Needles fitted with jeweled centers are sold by E. Goldbacher, 98 Fulton St., N.Y.

Construction of the Tangent Galvanometer. 49

a common sewing needle as a pivot, and if the magnetic needle is found not to balance so as to lie in a laterally horizontal position, adjust the center before the cement has firmly set. If one end of the needle appears heavier than the other, load the light end with a touch of melted sealing-wax, applied to the under side.

Procure a fine straight straw, $3\frac{1}{4}$ in. long, and make a transverse hole through the middle of it with a large pin ; thrust the top of the glass center carefully through this hole, so that the straw, which is to serve as an index or pointer, will lie exactly at right-angles with the magnetic needle (Fig. 37). Secure the straw to the glass center with a mere trace of glue, and set it away to dry.

Make a dial of card-board like Fig. 39 (which shows one half of it), one half being graduated in degrees and the other half divided

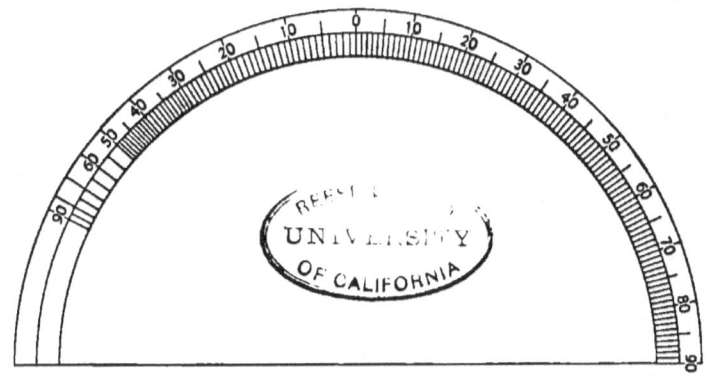

FIG. 39. Half of Card-board Dial of Tangent Galvanometer.

in tangents, upon the principle explained in (96). The outer circle should be $3\frac{3}{4}$ in. in diameter. Next cut out, in a lathe if possible, a circular hole 4 in. in diameter in a circular piece of hard wood H, 5 in. diameter and 1 in. thick, and then enlarge this hole $\frac{3}{16}$ in. all round with a shoulder $\frac{1}{4}$ in. deep, as shown in the cross-section, Fig. 36.

Make a bridge-piece D of hard wood, 1 in. thick, and secure it to the base by brass screws as shown in Fig. 35. This bridge-piece should be just high enough so that the center of the magnetic needle N S will come exactly in the geometrical center of the vertical ring C containing the wire. The circular board G is secured to the bridge-piece D with brass screws, the sewing-needle (point upward) inserted into it for a pivot, and the card-board dial secured thereto by small brass or copper tacks, in such position that when the magnetic needle is in the plane of the vertical ring, the straw will point to zero upon the scale at each end. The horizontal wooden ring H is now laid upon the dial and fastened in place with brass screws. A circular piece of glass *g* is cut to the right size to lie upon the shoulder of the ring H, and may be secured in place by an elastic ring *r* of stout brass wire, cut to the right length, so that it may be sprung into place over the glass, and within the wooden ring, after the needle is in place.

If pieces of mirror be let into the dial, they will materially aid in making accurate readings of the indications, as the index and its image will then appear coincident only when the eye is vertically over the point observed. The error arising from angular observations is termed *parallax*.

FIG. 40. Ordinary Binding-Screw.

FIG. 41. Binding-Screw, English Pattern.

The ends of the coil wire are carried through holes into grooves cut for the purpose on the under side of the base, and thence to two *binding-screws*, P N (English pattern), Fig. 35, which may be purchased for about 20 cts. each. These should be placed in the plane of the vertical ring upon opposite sides of the base, and the ends of the wires carefully soldered to them underneath the base.

Binding-screws are brass clamps for conveniently attaching connecting wires to electrical instruments. They are made in many patterns, the most common types being those shown in Figs. 40 and 41. In Fig. 40, the wire is inserted into the transverse hole and clamped by turning the screw. This form is very handy for ordinary purposes, but where a very good contact is essential, as in measuring apparatus, the English pattern, Fig. 41, is preferable, the wire being looped round the stem and clamped by the thumb-nut. For attaching two or more wires, this pattern is sometimes made with more than one nut, as in Fig. 42.

FIG. 42. Double Binding-Screw, English Pattern.

105. The accuracy of the above-described galvanometer will

Construction of the Rheostat.   51

depend largely upon the care used in making it. The most important point is to make sure that the centers of the vertical coil and of the magnetic needle exactly coincide, and that the wire is accurately and smoothly wound upon the vertical ring. It is possible for the amateur to construct an instrument which will do quite as accurate work as those which are sold for $50 and upward by professional instrument-makers (367).

106. Construction of the Rheostat.—A convenient form of rheostat for use with the tangent galvanometer is not beyond the constructive skill of the amateur. Procure from a dealer a device called a "peg pole-changer," Fig. 43, with 4 pegs, which may be had for about $3. This is to be mounted upon the cover,

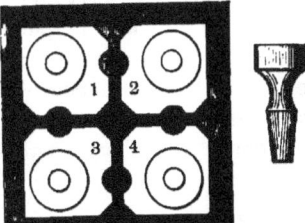

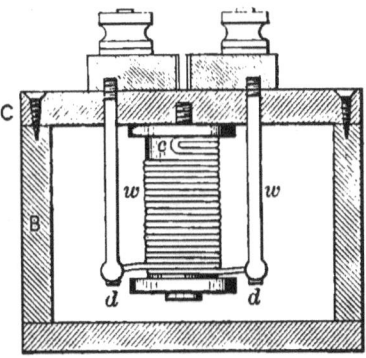

FIG. 43. Peg Pole-Changer and Peg.   FIG. 44. Rheostat and Box.

C, of a wooden box, B, which is removably fastened by screws, as shown in Fig. 44.

Procure also from a dealer two lengths of double cotton-covered German-silver resistance wire, as follows:

13 ft. of No. 18 (American gauge).........0403 in. diameter.
40 "   "   " 26   "   "   ....0159 in.   "

Stretch each piece of wire out straight, double it in the middle of its length, and wind smoothly upon a separate wooden spool,—a common thread spool will do. Commence winding at the bight or loop c, and wind double so that both ends of the wire will come on the outside of the coil, as at $d\,d$. Fasten the filled spools to the under side of the box cover, C, with long brass screws, as shown in Fig. 44.

The ends of the wires on the spools must be uncovered by scraping off the cotton envelope; then carefully cleaned with emery cloth, and soldered with rosin to stout brass wires $w\,w$, which are screwed into the brass plates supporting the binding-screws, precisely as indicated in Fig. 45. The spare peg-hole seen in Fig. 43 will provide for a third coil of still higher resistance (indicated in Fig. 45 in dotted lines at the right), in case one should be needed.[2]

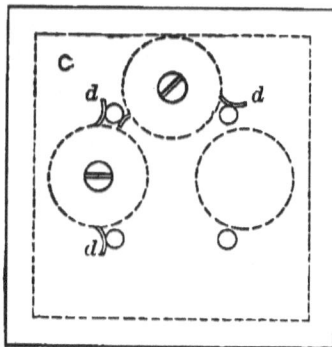

FIG. 45. Connections of Rheostat.

The thick wire coil, made as above, will have an approximate resistance of 1 ohm, and the thinner one an approximate resistance of 20 ohms (125).

107. Preparation for Experiments. — Having provided himself with the necessary appliances, the student is now prepared to undertake an experimental investigation of the laws of the electric current. Four gravity cells should be set up in accordance with directions in Chapter II, using pure water as directed in (21) and not $s. z.$ solution. Arrange these in some convenient place, as on a

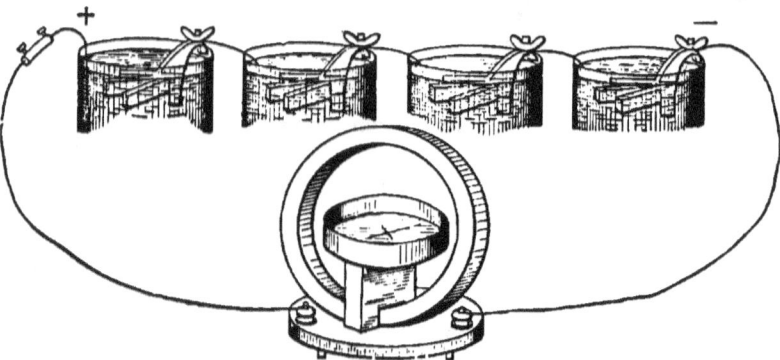

FIG. 46. Cells in Series with Tangent Galvanometer.

common kitchen table, and connect them with each other as in diagram Fig. 46, the copper of one cell to the zinc of the next, and

---

[2] These *resistance-coils*, before being mounted, may with advantage be dried for several hours in a hot, but not too hot, oven, and when taken out should be instantly immersed in a hot mixture in readiness for the purpose, composed of 10 parts by weight of rosin and 1 part of white wax. Let the compound cool with the coils in it, removing the latter just before it sets. Heat them again, if necessary, enough to remove the surplus material.

Effect of Varying Number of Cells in Series. 53

the free copper and zinc terminals, by means of pieces of copper wire about 3 ft. long, to the binding screws P N of the galvanometer. The cells are now said to be arranged in *series*, both with each other and with the galvanometer. At first, little or no effect will be observable upon the needle of the galvanometer; but in the course of a few hours a deflection may be observed, which will slowly increase. Leave the apparatus alone, except to tap the galvanometer occasionally with the finger to facilitate the movement of the needle, until it indicates about 58°, when the cells may be considered to be in good working condition.

108. **Effect of Varying Number of Cells in Series.**—Let the student now carefully observe the effect produced upon the needle of the galvanometer by varying in the number of cells in series with the coil. The results will be found to be something like the following:

Cells in Series.	Deflection.	Tangent.
4	58°	1.60
3	$57\frac{1}{4}°$	1.56
2	$56\frac{1}{4}°$	1.50
1	$53\frac{1}{2}°$	1.35

As elsewhere stated (96), the effective strength or quantity of current is always in the ratio of the tangents of the angles of deflection. Hence we have, in the above experiment, the apparently paradoxical result, which is nevertheless susceptible of rational explanation (132), that under the conditions stated, quadrupling the number of the cells in the circuit only increases the quantity of current in the ratio of 135 to 160, or about 18 per cent.

109. **Cells in Parallel Series.**—Next arrange the four cells in 2 series, with 2 cells in each series, as in diagram, Fig. 47. Such

FIG. 47. Cells in Parallel Series.

an arrangement is termed *parallel series*, or sometimes *multiple series* This gives quite a different result.

2 series of 2 cells each. Deflection $69\frac{3}{4}°$ (tangent 2.70).

110. **Cells in Parallel.**—Finally, connect all the copper terminals to one terminal of the galvanometer and all the zincs to the other, as in Fig. 48. This is termed connecting in *parallel*, or *multiple-arc*. The result will be again different, as follows:

4 cells in parallel, deflection 73¼° (tangent 3.38).

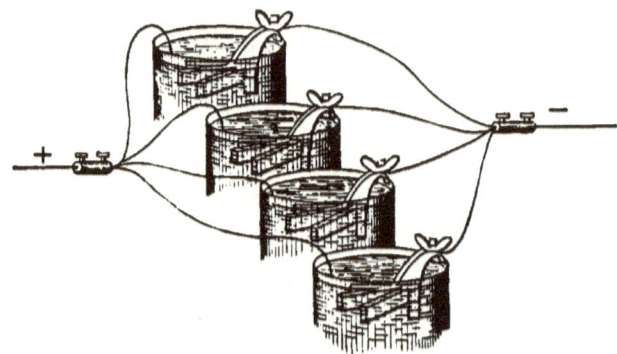

FIG. 48. Cells in Parallel.

111. **Increasing Length of Conducting Circuit.**—Leaving the cells and the galvanometer connected in the manner last described, let the current next be made to pass also through the length of the 2 lbs. of No. 18 copper wire, which will be in linear measure a little over 300 feet. The deflection of the needle will now fall from 73¼° to about 59¼° (tangent 1.70). This experiment proves that when the current is made to pass through the long copper wire, under the conditions of the experiment, its quantity is diminished to about one-half the original amount.

112. Next place the galvanometer in circuit with one cell, as we did once before, and found the deflection to be 53¼° (tangent 1.35). Insert the long copper wire in circuit with the galvanometer and cell, and we get a deflection of about 44° (tangent 0.965), showing that the quantity of current has been diminished about one-third.

113. Leaving the long copper wire still in circuit with the galvanometer, add another cell, making 2 in series. The deflection now becomes about 51° (tangent 1.23), and by adding still another cell, making 3, we get a deflection of 53¼° (tangent 1.35), exactly what we had with one cell when the copper wire was not included in the circuit.

114. **Conditions which Determine Quantity of Current.**—From the last experiment we arrive at two fundamental facts re-

Table of Tangents. 55

TABLE III.
TANGENTS FOR EVERY HALF DEGREE.

Deg.	0′	30′	Deg.	0′	30′	Deg.	0′	30′
0	.0000	.0087	30	.5774	.5890	60	1.732	1.767
1	.0175	.0262	31	.6009	.6128	61	1.804	1.842
2	.0349	.0437	32	.6249	.6371	62	1.881	1.921
3	.0524	.0612	33	.6494	.6619	63	1.963	2.006
4	.0699	.0787	34	.6745	.6873	64	2.050	2.096
5	.0875	.0963	35	.7002	.7133	65	2.144	2.194
6	.1051	.1139	36	.7265	.7400	66	2.246	2.300
7	.1228	.1317	37	.7536	.7673	67	2.356	2.414
8	.1405	.1495	38	.7813	.7954	68	2.475	2.539
9	.1584	.1673	39	.8098	.8243	69	2.605	2.675
10	.1763	.1853	40	.8391	.8541	70	2.747	2.824
11	.1944	.2035	41	.8693	.8847	71	2.904	2.989
12	.2126	.2217	42	.9004	.9163	72	3.078	3.172
13	.2309	.2401	43	.9325	.9490	73	3.271	3.376
14	.2493	.2586	44	.9657	.9827	74	3.487	3.606
15	.2679	.2773	45	1.000	1.018	75	3.732	3.867
16	.2867	.2962	46	1.035	1.054	76	4.011	4.165
17	.3057	.3153	47	1.072	1.091	77	4.331	4.510
18	.3249	.3346	48	1.111	1.130	78	4.705	4.915
19	.3443	.3541	49	1.150	1.171	79	5.145	5.395
20	.3640	.3739	50	1.192	1.213	80	5.671	5.976
21	.3839	.3939	51	1.235	1.257	81	6.314	6.691
22	.4040	.4142	52	1.280	1.303	82	7.115	7.599
23	.4245	.4348	53	1.327	1.351	83	8.144	8.777
24	.4452	.4557	54	1.376	1.402	84	9.514	10.39
25	.4663	.4770	55	1.428	1.455	85	11.43	12.71
26	.4877	.4986	56	1.483	1.511	86	14.30	16.35
27	.5095	.5206	57	1.540	1.570	87	19.08	22.90
28	.5317	.5430	58	1.600	1.632	88	28.64	38.19
29	.5543	.5658	59	1.664	1.698	89	57.29	114.6

specting the electric current, which are that its *quantity* may be affected,—*first*, by varying the length of the conductor which it traverses, and *second*, by varying the number of cells in series in the battery from which the current is derived.

115. **Resistance.**—What is true of the copper wire is also true of all known substances,—namely, that they oppose a certain and definite *resistance* to the passage through them of an electric current, and the quantity of current which passes, other things being equal, is in every case *inversely proportional to the resistance* which it encounters in the circuit (127). In other words, the greater the resistance the less the quantity of current, and *vice versa*.

116. **Conductors and Insulators.**—Although all bodies offer more or less resistance to the passage of the electric current, there is an enormous difference in the resisting capacity of different substances. Those which offer comparatively little resistance are called in a general sense *conductors* of electricity, while those that offer great resistance are termed *insulators*. This distinction, like that, for example, between heat and cold, is wholly relative and not absolute. The most perfect known conductors offer some resistance to the current, and the most perfect insulators known permit some current to pass. But the actual difference in some instances is almost beyond the power of the mind to grasp.

> It is difficult to find any comparison which will give a tolerably good idea of the extraordinary difference between the electrical resistance of these two materials (copper and gutta-percha). It is about as great as the difference between the velocity of light and that of a body moving through one foot in 6700 years ; yet the measurements of the two quantities are daily made with the same apparatus and the same standards of comparison. This fact is well calculated to give an idea of the range of electrical measurements, and the perfection to which the instruments employed have been brought.— FLEEMING JENKIN on Submarine Telegraphy, in *North British Review*, December, 1866.

117. The division of bodies into the two classes of conductors and insulators, though in a certain sense arbitrary, is very convenient in practice. In telegraphy, the term conductor is applied to all substances which are used in any manner as a portion of the conducting circuit, while on the other hand, the term insulator is applied to all substances which are employed to confine the electrical current to such conductors, by preventing its escape in undesired directions. Following is a list of some of the substances so used, arranged as nearly as possible in the order of their specific conductivity :

Specific Resistance of Different Metals. 57

CONDUCTORS.

1. Copper.
2. Zinc.
3. Platinum.
4. Iron.
5. Nickel.
6. Tin.
7. Lead.
8. Mercury.
9. Carbon.
10. Acids.
11. Aqueous solutions of metallic salts.
12. Water.
13. Wood (moist).
14. Earth (moist).

INSULATORS.

15. Metallic oxides.
16. Ice (dry).
17. Paper (dry).
18. Wood (dry).
19. Earth (dry).
20. Cotton.
21. Glazed porcelain.
22. Silk.
23. Bitumen.
24. Sulphur.
25. Oxydized oils.
26. Balata.
27. India-rubber.
28. Gutta-percha.
29. Shellac.
30. Paraffin.
31. Hard Rubber.
32. Glass.
33. Air.

The conducting power of all alloys or mixtures of different metals is very much less than that of any one of the metals of which they are composed.[2] The air is the most perfect non-conductor known, even when charged to saturation with aqueous vapor, but it should be remarked that when the moisture of such vapor is deposited upon the surface of insulating supports, it may form a conducting film of water.

117a. **Specific Resistance of Different Metals.**—The resistance referred to in (115), is *specific resistance*, a quality which depends, in some way not definitely understood, upon the internal molecular structure of each particular substance. Thus an iron wire of the same weight and thickness as the copper wire used in the preceding experiments (111. 112) would offer nearly 6 times as much resistance.

If the resistance of a pure copper wire of a certain length and diameter is known, the resistance of other wires, of similar dimensions but of other metals, may be found by the following rule:

For Brass, multiply copper resistance by............ 4.5
" German-silver   "   "   " 12.9
" Iron   "   "   " 5.9
" Platinoid   "   "   " 19.5
" Platinum   "   "   " 14.8

[1] For Matthiessen's determinations of the specific resistance of metals and alloys, see SPRAGUE : *Electricity*, etc., p. 282 ; see also BENOIT: *Jour. Soc. Tel. Eng.* iv. 112.

118. Conditions Affecting Resistance.—In the case of a body having a certain specific resistance, the actual measurable resistance depends upon certain conditions, viz.:

(i) **Temperature.**—The resistance of all metals increases with the temperature. The converse of this is true of most liquids at temperatures above the freezing point (162).

(ii) **Length.**—The resistance of any body is in proportion to its length, measured in the direction traversed by the current.

(iii) **Cross-section.**—The resistance of any body is inversely in proportion to its area of its cross-section. Doubling the cross-section halves the resistance.

Every resistance capable of being measured, must necessarily be equal to the resistance of a certain length of a standard conductor; therefore *resistance may be expressed in terms of length.* Thus a telegraph line made up of two or more sections of wire, of unequal thickness or gauge, and so presenting a different resistance per mile of length, may be expressed in terms of the resistance of a given length of a standard wire: that is to say, the actual resistance of any telegraph line must represent and be equal to the resistance of a certain length of wire of the standard gauge. This length is called the *reduced length* of the line. The convenience of this mode of reducing resistance to terms of length in making tests and measurements of lines will appear hereafter.

119. Provisional Theory of Electricity.—Before undertaking to analyze and explain the results which have been observed in the foregoing experiments, it will be convenient for the student to form some sort of a mental conception of the agency which produces the phenomena observed. A distinguished electrical engineer has remarked:

The student of electricity, in considering the various phenomena which come under his notice, must of necessity form some theory in his mind as to the nature of the element with which he has to deal; and as philosophers are not in accord as to its nature and the theory of its action, the choice must to a novice be a difficult one. Without, therefore, in the least offering any opinion on this point, I would advise him, until his ideas are more matured, to regard electricity as a *substance* like water or gas, having a veritable existence, and also easily converted into heat and, *vice versa*, in other respects indestructible. LATIMER CLARK: *Electrical Measurement*, p. vii.

120. Mechanical Analogue of Electrical Action.—According to this manner of looking at the question, we may regard a voltaic cell or a dynamo-electric machine as a sort of pump, by which positive electricity is pumped through an endless channel or

Conception of Potential and Electromotive Force. 59

conduit, as water might be pumped through a pipe to the top of a hill, and thence allowed to flow back in a definite channel to the foot of the pump from whence it started. This illustration, if fixed in the mind, will materially aid the student in forming a useful mental conception of the character of the flow of electricity in a circuit of conductors.

121. **Conception of Potential and Electromotive Force.** —When water is pumped to an elevation by the application of power and then allowed to run back to its original level, it is capable of being *made to do work* in the course of its descent by means of water-wheels or otherwise, and it is obvious that the *amount of work* it is capable of doing under these conditions depends upon three things: (1) the *height* of the fall, (2) the *quantity* of water, and (3) the length of *time* the effect continues. We may express the condition of affairs by saying that the water which has been raised has a certain *potential energy*, which may be defined as *capacity to do work*, and for brevity we may call it *potential*. We may, therefore, for present purposes, regard electricity as a material fluid to which a certain potential has been given by the action of a battery or of a magneto-electric machine (80), and which is therefore capable of doing mechanical work, as is the case with the descending water. That quality of a voltaic cell, or of a magneto or dynamo-electric machine, by virtue of which it confers potential upon electricity is termed *electromotive force*, usually abbreviated to *e. m. f.*

122. **Practical Electric Units.**—The units of force and work, and their relation to force of gravitation, have already been referred to (91), and it has also been explained (94) that a definite quantitive relation exists between mechanical force and the force of an electric current. This fact enables us to base our practical units of electrical measurement upon absolute units; or in other words, *our practical units are derived from constants furnished by nature*. The relation or law connecting the two forces is as follows:

The force which a given current traversing a circular arc exercises upon a magnetic pole of given strength situated at its center, is equal to the strength of the pole multiplied by the strength of the current and also by the length of the arc, and divided by the square of the radius or semi-diameter; that is to say, the distance from the wire from the magnet pole.

The names which have been given to the practical electrical units are derived from the names of philosophers of various nationalities who have distinguished themselves by electrical discoveries and investigations.

123. The Ampère.—The *unit of current* is called the ampère.[1] It is equivalent in value to $\frac{1}{10}$ of the absolute or *c. g. s.* unit of current (91). The actual value of any current within the range of the instrument may be determined from the indications of a properly constructed tangent galvanometer. This is done by the aid of the following rules:

(i) Given the number of turns and the mean radius of the coil of the tangent galvanometer, the observed deflection, and the horizontal force of the earth's magnetism at the place of observation (94), to find the current in ampères:

RULE.—*Multiply together the mean radius in inches, the tangent of the deflection, the horizontal magnetic intensity of the earth in dynes, and the reduction factor 4.0425, and divide the product by the number of turns in the coil. The quotient is the current in ampères.*

Example.—In a tangent galvanometer having a 10-in. coil wound in 6 turns, and giving a deflection of 60° in Washington, D. C., in 1885, required the value of the current.

Ans.—5 (in.) × 1.732 (tan. of 60°) × .2026 (dynes) × 4.0425 (reduction factor) ÷ 6 (turns) = 1.182 ampères.[4]

(ii) Given the number of turns and mean radius of the coil, the value of the current and the horizontal intensity, to find the deflection.

RULE.—*Multiply together the number of ampères and the number of turns, and divide this product by the product of the mean radius in inches, the horizontal intensity in dynes, and the reduction factor 4.0425. The quotient will be the tangent of the deflection.*

Example.—A current of 0.25 ampères was passed through the coil of the above described galvanometer in New York. What was the deflection?

Ans.—0.25 (ampères) × 6 (turns) ÷ [5 (in.) × 0.1872 (dynes) × 4.0425 (red. fac.)] = 0.3874 = tangent of 21° nearly.

---

[1] AMPÈRE (ANDRE MARIE), an eminent French philosopher and mathematician, in honor of whom the unit of current received its name; born at Lyons, 1775. He became inspector-general of the university (1808); professor in the Polytechnic School (1809); and Member of the Institute (1814). Having made important discoveries in electro-magnetism, he published (1822) his *Collection of Observations on Electro-Dynamics*, a remarkable work. "The vast field of physical science," says Arago, "perhaps never presented so brilliant a discovery, conceived, verified, and completed with such rapidity." He subsequently published his *Theory of Electro-Dynamic Phenomena, Deduced from Experiments* (1826). Died in Marseilles, 1836.

[4] Each unit in the metric system has its decimal multiples and sub-multiples; that is to say, measures larger or smaller than the standard unit. These multiples and sub-multiples are denoted by prefixes, derived from the Greek and Latin languages respectively, placed before the names of the units, as follows:

PREFIX.	SIGNIFICATION.	PREFIX.	SIGNIFICATION.
Micro-	means one millionth of a ——.	Hecto-	means hundred times a ——.
Milli-	" one thousandth of a ——.	Kilo-	" thousand times a ——.
Centi-	" one hundredth of a ——.	Myria-	" ten thousand times a ——.
Deci-	" one tenth of a ——.	Mega-	" one million times a ——.
Deka-	" ten times a ——.		

Thus a millimetre is one-thousandth of a metre; a milliampère is one-thousandth of an ampère; a megohm is one million ohms, etc., etc.

The Coulomb.—The Volt.—The Ohm.

(iii) *Example.*—The earth's magnetism in Toronto, Ont., being the directing force of a galvanometer having a 10-in. coil, how many turns must be put on it in order that a current of 0.96 ampères shall give a deflection of 60°? *Ans.*—6 turns.

From the above explanations and examples, it will be seen that a standard galvanometer may be constructed, from which it is always possible, knowing the force of the earth's magnetism, to determine the value of any electric current in ampères. In some cases, tangent galvanometers are graduated so that the ampères may be read directly without calculation. Such an instrument is called an *ampère-meter*, or more commonly, an *ammeter* (369).

123a. **The Coulomb.**—The quantity of current which traverses a circuit in 1 second, when the strength of current is 1 ampère, is termed a Coulomb.[5] It is a unit which is of little or no practical utility in ordinary telegraph work, or in fact for any other purpose, and is referred to here only because it has been given a place in the accepted system of electric units.

124. **The Volt.**—The unit of *electromotive force* is called the volt.[6] It closely approximates that of a single sulphate of copper or gravity cell in good condition (24), so that in telegraphic work it is usually accurate enough for practical purposes to estimate 1 cell equals 1 volt. Accurately, 1 gravity cell has an *e. m. f.* of 1.07 volts. This value is subject to slight variation from various causes. It is not much influenced by temperature (153, note).

125. **The Ohm.**—The unit of electrical *resistance* is called the ohm.[7] It is equal to the resistance to a column of pure mercury, 1 sq. millimetre in cross-section and 106 centimetres (more or less), in length,[8] at a temperature of 0° Centigrade or 32° Fahrenheit.

[5] COULOMB (CHARLES AUGUSTIN DE), a distinguished mathematician, born at Angoulême, France, 1736. He is regarded as the founder of experimental physics in France. The theory of electricity is largely indebted to the investigations of this philosopher. Died, 1806.

[6] VOLTA (ALESSANDRO), born at Como, Italy, 1745; was first professor of physics at Como, and afterward in the University of Pavia, where he taught and studied for 30 years. In 1782, he invented the electrical condenser (317), and finally arrived at the invention of the famous cell which bears his name (8), which he described in a letter to Sir Joseph Banks in 1800. Summoned to Paris by Napoleon, he received the gold medal of the Institute, of which he became a member in 1802. His works were published in 9 volumes, in Florence, in 1816. Died, 1827.

[7] OHM (GEORG SIMON), born at Erlangen, Bavaria, 1787; studied in his native city; was appointed (1817) professor of physics at the Jesuit college of Cologne, director of Polytechnic School at Nuremburg (1883), and professor (1849) at Munich, where he died in 1874. He discovered the so-called Ohm's law (124), which he published in 1827, and for which was awarded the Copley medal by the Royal Society of London.

[8] In 1861, the British Association for the Advancement of Science, at the suggestion

62 Laws and Conditions of Electrical Action.

The resistances of various wires used as telegraphic conductors is given in the tables, pp. 94, 112.

126. Resistance of Liquids.—The resistance of liquids is enormously greater than that of metallic substances. The relative specific resistances of some of the voltaic solutions used in telegraphy are as follows: [9]

Pure copper (standard of comparison).................1.
Pure rain water.......................................40,653,723.
Water 12 parts, sulphuric acid 1 part.................. 1,305,467.
Sulphate of copper 1 lb., water 1 gallon...............18,450,000.
Saturated solution sulphate of zinc....................17,330,000.

Table iv, on the next page, contains the results of more recent determinations of the specific resistance of copper and zinc solutions at various temperatures, computed from the experiments of Becker.[10] As the temperature rises, the resistance falls off. This effect is further referred to in (164).

of Sir William Thomson, appointed a committee on electrical standards, which, after a long series of experiments by eminent physicists, determined the value of the ohm to be nearly that of a column of pure mercury 105 centimetres long and 1 square millimetre in cross-section, at temp. 0° centigrade, and officially caused resistance coils made of wire of an alloy of platinum and silver to be issued as standards. Resistance coils copied from these standards are known as *B. A. units* or *ohms*. More recent careful determinations of Lord Rayleigh and many others have proved beyond doubt that the B. A. unit or ohm is more than 1 per cent. too small. An attempt has accordingly been made to substitute for the old standards new ones of the corrected value. In accordance with the recommendation of the International Congress of Electricians, held in Paris in 1884, a *legal ohm* is defined to be a mercury column of the above section, and 106 cm. in length. The exact ratio is:

1 Legal ohm = 1.0112 B. A. ohm.
1 B. A. ohm = 0.9889 legal ohm.

At the meeting of the British Association in September, 1890, it was recommended that the value for the mercury column of 106.3 cm. be substituted for the 106 cm. of the International Congress, and it is not unlikely that this value may ultimately be adopted.

In Germany, the *Siemens unit* (known as the S. U.) is largely used, and many of the older instruments now in use in the United States are adjusted to this standard. It is designed to be equal to a column of mercury 1 metre long and 1 sq. mm. cross-section at temp. 0° C.

1 Siemens unit = 0.9540 B. A. ohm.
1 B. A. ohm. = 1.0486 S. U.

[9] MOSES G. FARMER: Shaffner's *Telegraph Manual*, 514.

[10] F. JENKIN: *Electricity and Magnetism*, 259. The maximum conductivity of *s. z.* solution is 23.5 per cent (s. g. 1.286) according to KOHLRAUSCH: *Physical Measurement*, p. 326. For other tables of resistances of liquids, see SPRAGUE: *Electricity*, etc. (2d ed.), 298; STEWART and GEE: *Elementary Practical Physics*, 219; NIAUDET: *Electric Batteries* (Fishback's Translation), 255; PRESCOTT: *Electricity and Elec. Tel.*, 182. For method of measurement see F. KOHLRAUSCH: *Jour. Soc. Tel. Eng.* xiii, 290.

TABLE IV.
SPECIFIC RESISTANCES OF VOLTAIC SOLUTIONS.
SULPHATE OF COPPER.

Percentage of salt in Solution.	14°	16°	18°	20°	24°	28°	30°	Centigrade.
8	45.7	43.7	41.9	40.2	37.1	34.2	32.9	Resistance of 1 cubic centimetre expressed in ohms.
12	36.3	34.9	33.5	32.2	29.9	27.9	27.0	
16	31.2	30.0	28.9	27.9	26.1	24.6	24.0	
20	28.5	27.5	26.5	25.6	24.1	22.7	22.2	
24	26.9	25.9	24.8	23.9	22.2	20.7	20.0	
28	24.7	23.4	22.1	21.0	18.8	16.9	16.0	

SULPHATE OF ZINC.

	10°	12°	14°	16°	18°	20°	22°	24°	Centigrade.
96 grams in 100 c. c. of solution.	22.7	21.4	20.2	19.2	18.1	17.1	16.3	15.6	Resistance of 1 cubic centimetre expressed in ohms.
Same solution with an equal volume of water.				21.1	20.3	19.5	18.8	18.1	17.3	

127. **Ohm's Law.**—The fundamental relation which exists in every electric circuit between electromotive force, resistance and current, is expressed by Ohm's law, which may be formulated in the following propositions :

(i) In any electric circuit, the *current* is the quotient of the electromotive force divided by the resistance ; hence the current in ampères may be found by dividing the e. m. f. in volts by the resistance in ohms.

(ii) In any electric circuit, the *electromotive force* is the product of the current and the resistance ; hence the total e. m. f. in volts may be found by multiplying together the current in ampères and the resistance in ohms.

(iii) In any electric circuit, the *resistance* is the quotient of the electromotive force divided by the current ; hence the resistance in ohms may be found by dividing the e. m. f. in volts by the current in ampères.

128. **Joule's Law.** [11]—The relation which exists between current and mechanical work is expressed by Joule's law, which may be formulated in the following propositions:

[11] JOULE (JAMES PRESCOTT), born in Salford, England, 1818. A self-taught philosopher, distinguished for the extent, originality, and accuracy of his physical researches. He ascertained in 1841, the law of the evolution of heat by the electric current (128), and determined in 1850, the numerical ratio of equivalency between heat and mechanical force (92). His discoveries, which are too numerous to permit more than general

64 Laws and Conditions of Electrical Action.

(iv) In any electric circuit, the *rate of doing work* is the product of the *e. m. f.* and the current; hence the rate in which work is being done in watts (150) may be found by multiplying together the *e. m. f.* in volts and the current in ampères.

(v) In any electric circuit, the *rate of doing work* is the product of the current multiplied into itself and into the resistance; hence rate of working in watts may also be found by multiplying together the resistance and the square of the current in ampères.

129. The term *work*, as herein used, includes the work which appears in the form of heat, as well as that which produces physical motion.

130. **Experimental Proof of Ohm's Law.**—The student is now prepared to understand an explanation of the results which have been referred to in the preceding paragraphs (108—113). Referring to the diagram, Fig. 49, we may trace the circuit as follows: Beginning at the copper or the positive pole of the battery, thence through the 62 feet of copper wire which forms the coil of the galvanometer; thence to the zinc Z or negative pole of the battery; thence in succession through the *s. z.* solution and the *s. c.* solution S to the copper plate C of the first cell; thence to the zinc plate Z of the next cell, and through the solution to the copper, and so on through the series until the starting point, the copper plate of the terminal cell, is reached.

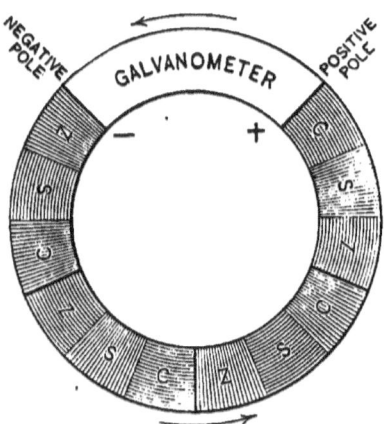

FIG. 49. Diagram of Galvanometer and Battery Circuit.

131. **Internal Resistance of the Cell.**—The solution in each cell may be regarded as a liquid conductor of cylindrical form, having a length of about 3 in. (the average distance between the copper and zinc plates) and a cross-section of about 28 sq. in. When a cell is in good working condition, the resistance of the contained liquids, at ordinary temperatures, is about 4 ohms, and may be regarded as

mention here, have been intimately related to the remarkable theory of the correlation of the physical forces (p. 38, note 8) which was developed by Mayer, Helmholtz, Seguin, Faraday, and Grove. His researches in electro-magnetism, particularly in respect to its application as a motive power, were extensive and important. Honors were conferred on him by almost every learned society in the world. His scientific papers were collected and published by the Physical Society of London in 1884. Died 1889.

Internal Resistance of the Cell.—First Case. 65

approximately equivalent to that of 250 feet of copper wire of the thickness of that in the coil of our galvanometer (103). The actual resistance of the zinc and copper plates of each cell, being at most but an insignificant fraction of 1 ohm, may in the present instance be disregarded in our computations.

132. First Case.—In the first example, we begin with 4 cells in circuit in a single series, as shown in Fig. 50.

This figure is a diagrammatical or conventional illustration of precisely the same thing which is shown in Fig. 46. The zinc plate of each cell is represented by a thick black line, and the copper by a thin line. The symbol for the galvanometer explains itself. In like manner, Fig. 51 corresponds to Fig. 47, and Fig. 52 to Fig. 48. In Figs. 51 and 52, a black dot at the intersection of two wires indicates that they are electrically united at the junction. This conventional representation of batteries, galvanometers, circuits, and other appliances will be employed hereafter in this work (208).

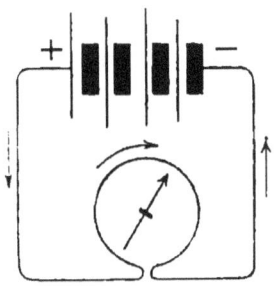

FIG. 50. Cells in Series with Galvanometer.

In this case, each cell has an approximate $e.m.f.$ of 1 volt (124), this value depending not at all upon the size of the element, but solely upon its chemical constitution. The aggregate $e.m.f.$ of the 4 cells of the series is therefore 4 volts. The aggregate resistance of the 4 cells is 16 ohms, and that of the galvanometer 1 ohm. We may neglect also the inappreciable resistance of the short connecting wires between the battery and the galvanometer, and call the sum of the resistances in the circuit (battery and galvanometer) 17 ohms. By Ohm's law (127, 1), we divide the $e.m.f.$ 4 (volts) by the resistance 17 (ohms), and our quotient is 0.235 (ampères).

With 3 cells in like manner, we have an $e.m.f.$ of 3 volts, an aggregate resistance of 13 ohms, and by Ohm's law a current of 0.230 ampères; and so in the remaining cases. Continuing this method of procedure, we get results which may be tabulated as follows:

Cells in Series.	Deflections.	Tangents.	E. M. F.	Resistance.	Current.
4	58°	1.60	4	.17	0.235
3	57¼°	1.56	3	.13	0.230
2	56¼°	1.50	2	.9	0.222
1	53½°	1.35	1	.5	0.200

We find, therefore, that the tangents of the angles of deflection are in proportion to the strength of the current in ampères, as computed by Ohm's law from known electromotive forces and known resistances.

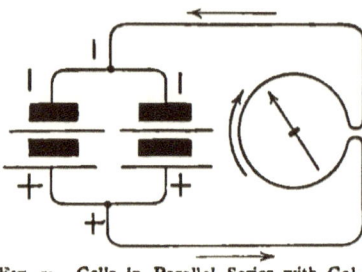

Fig. 51. Cells in Parallel Series with Galvanometer.

133. **Second Case.**—Take the next arrangement (109), in which we have 2 series of cells and 2 cells in each series, Fig. 51. The question now arises: If the resistance of each cell is 4 ohms, what will be the resistance of the group? It is *less* than in the preceding case, as the increased deflection of the needle shows, and as might have been inferred from the fact that the current from each series of cells does not now pass through the other series, nor encounter its resistance. Neither is the *e. m. f.* of one series superimposed upon that of the other series as before. A little reflection will make it clear that the present arrangement is precisely equivalent to 2 cells in series, each having copper and zinc plates of double the original area. Hence we may consider the cross-section of the liquid conductor to be doubled, while its length remains unaltered, from which it follows that its resistance is but half what it was originally (118).

134. **Law of Joint Resistances.**—The law determining the resistance of any circuit which divides into two or more branches which reunite at another point, is a general one, and applicable in all such cases, whether of batteries or of conductors. The resistance offered by two or more such branches is termed their *joint resistance*, and is computed by the following rules:

RULE 1.—*Add together the reciprocals of the individual resistances of all the branches, and the reciprocal of the result will be the joint resistance of the group.*

The *reciprocal* of any number is the fraction obtained by dividing unity (or 1) by that number; and the reciprocal of any common fraction, is that fraction itself inverted. Thus the reciprocal of 2 is ½ or 0.5; and conversely, the reciprocal of 0.5 or ½ is 2. The reciprocal of ¾ is 4/3. A table of reciprocals is given on page 67.

In case there are only two branches, a simpler method of computation may be used:

RULE 2.—*Multiply together the individual resistances of the two branches, and divide the product by their sum; the quotient will be the joint resistance.*

Table of Reciprocals.

TABLE V.

RECIPROCALS OF NUMBERS FROM 1 TO 100.

No.	Rec.	No.	Rec.	No.	Rec.	No.	Rec.	No.	Rec.
1	1.000	21	.0467	41	.0244	61	.0164	81	.0123
2	.5000	22	.0454	42	.0238	62	.0161	82	.0122
3	.3333	23	.0434	43	.0232	63	.0159	83	.0120
4	.2500	24	.0416	44	.0227	64	.0156	84	.0119
5	.2000	25	.0400	45	.0222	65	.0154	85	.0118
6	.1667	26	.0385	46	.0217	66	.0151	86	.0116
7	.1428	27	.0370	47	.0213	67	.0149	87	.0115
8	.1250	28	.0357	48	.0208	68	.0147	88	.0114
9	.1111	29	.0344	49	.0204	69	.1045	89	.0112
10	.1000	30	.0333	50	.0200	70	.0143	90	.0111
11	.0909	31	.0323	51	.0196	71	.0141	91	.0110
12	.0833	32	.0312	52	.0192	72	.0139	92	.0108
13	.0769	33	.0303	53	.0188	73	.0137	93	.0107
14	.0714	34	.0294	54	.0185	74	.0135	94	.0106
15	.0667	35	.0286	55	.0182	75	.0133	95	.0105
16	.0625	36	.0277	56	.0178	76	.0131	96	.0104
17	.0588	37	.0270	57	.0175	77	.0130	97	.0103
18	.0555	38	.0263	58	.0172	78	.0128	98	.0102
19	.0526	39	.0256	59	.0169	79	.0126	99	.0101
20	.0500	40	.0250	60	.0166	80	.0125	100	.0100

Any sum *multiplied* by the reciprocal of a number is equal to the same sum *divided* by the number corresponding to the reciprocal. In the table, the reciprocals are those of whole numbers, but it is easy to extend their use to decimals, or to mixed numbers, by shifting the decimal point; thus, the

$$
\begin{aligned}
\text{Reciprocal of } 390 &= .00256 \\
\text{`` `` } 39 &= .0256 \\
\text{`` `` } 3.9 &= .256 \\
\text{`` `` } .39 &= 2.56 \\
\text{`` `` } .039 &= 25.6
\end{aligned}
$$

68 *Laws and Conditions of Electrical Action.*

135. In the present case we have 2 branches, with a resistance of 8 ohms in each branch. Hence we have $8 \times 8 = 64$; $8+8 = 16$; $64 \div 16 = 4$; add galvanometer 1, and we have as total resistance 5. Dividing the *e. m. f.*, 2, by this amount gives a current of 0.4 ampères. We have therefore:

Cells in parallel series.	Deflection.	Tangent.	E. M. F.	Resistance.	Current.
4	69⅞°	2.70	2	5	0.4

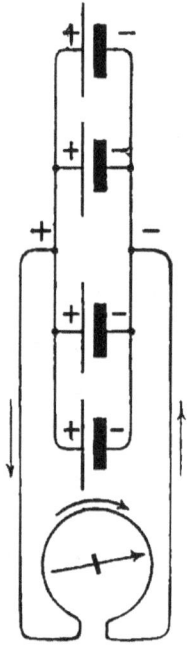

Fig. 52. Cells in Parallel with Galvanometer.

136. **Third Case.**—Next we have (110) the 4 copper terminals connected to one terminal of the galvanometer and the 4 zincs to the other terminal, as in Fig. 52. In this case, by the rule (134), the reciprocal of 4 is 0.25; the sum of the four reciprocals is therefore 1, the reciprocal of which is 1, and this added to the galvanometer resistance 1, makes a total of 2, while the *e. m. f.* is now reduced to 1. Hence, we have in this case:

Cells in series.	Deflection.	Tangent.	E. M. F.	Resistance.	Current.
1	73½	3.38	1	2	0.5

137. Passing next to the experiment in (108), in which we found that having 3 cells in circuit with the galvanometer, and 300 feet of a certain gauge copper wire, the deflection apparently indicated that we produced exactly the same current in the circuit that we did with 1 cell when the copper wire was not included. Let us see whether Ohm's law accounts for the result. We find from the copper wire table (p. 94) that the resistance of the length of wire included is approximately 2 ohms. We have, therefore:

Resistance of 3 cells battery ....12 ohms.
"   " galvanometer................. 1   "
"   " 300 feet copper wire 2   "

Total.......................15 ohms.

3 (volts) ÷ 15 (ohms) = 0.2 (ampères).

Branch or Derived Circuits.   69

In the other case we had :

  Resistance of 1 cell .... 4 ohms.
   " " galvanometer .... 1 "

  Total .... 5 ohms.

  1 (volt) ÷ 5 (ohms) = 0.2 (ampères).

  138. Ohm's law is therefore confirmed in every particular by the results of experiment, and observation, and we learn, moreover, the important fact that *the quantity of current traversing any given circuit may be varied either by varying the electromotive force or by varying the resistance.*

  139. We also learn from Ohm's law, as interpreted by the experiments which have been made, that *every portion of an undivided or non-branching circuit is traversed by the same quantity, or number of ampères, of current at the same time, without reference to its relative resistance.*

  140. **Currents in Branch Circuits.**—When any circuit divides into two or more branches a current traversing that circuit distributes itself between these branches inversely in proportion to their respective resistances, or, what is the same thing, directly in proportion to their several conductivities. The branches are also termed *shunts* or *derived circuits*. Each such branch may be regarded as a shunt to all the other branches in parallel with it.

  The word shunt is of English origin, and is derived from the analogy of a railroad siding where trains pass each other, which in that country is known as a shunt.

  141. **Electric Potential.**—Having thus gained some experimental as well as theoretical knowledge of electromotive force (121), resistance (115), and current (91), the student should next endeavor to acquire a definite understanding of the meaning of the term *potential*. The resemblance between the behavior of electricity and that of a material fluid like water has already been pointed out (120). Recurring to this analogy, if we assume a stream of water to be flowing through a closed pipe, we know that as soon as the flow has become steady, exactly the same number of gallons per minute will pass through every cross-section of the pipe, whatever may be the difference in its diameter at different points. This is exactly analagous to that which occurs in the case of an electric current (139).

  142. Although the quantity of water which passes must necessarily be the same in every cross-section of the pipe, the pressure

per square inch is by no means equal throughout, and this is true whether the pipe is level and whether it is of uniform diameter or otherwise. As we proceed along a horizontal pipe in the direction of the flow, we observe the pressure becomes less and less as we go farther away from the supplying reservoir.

143. Illustration of Fall of Potential.—A like effect takes place in the electric conductor. A fall, technically termed a *drop in potential*, occurs as we recede from the source of electricity, just as there is a fall of pressure in the water-pipe. For example, let Fig. 53 represent a vessel filled with water.[12] The tap at C is closed, and the water stands at the same level in all the vertical tubes, showing that no difference of pressure exists, and consequently there can be no current of flow in the liquid. But when the tap at C is opened, as in Fig. 54, it will be observed that the level in the several vertical tubes stands lower and lower as we pass from A toward C. The height of water in each tube indicates the pressure which exists at the point of its junction with the tube B. This difference in hydrostatic pressure between different points in the pipe produces the flow of water which we call a current. The original cause of the flow is manifestly the force which lifted the water in the first place to a point above the level of the pipe B, and thus conferred upon it the pressure or potential which it now has (121). Therefore we may say without error, that *electromotive force causes potential to exist.* When resistance is removed, a fall of potential occurs at some point, and this fall of potential gives rise to an electric current. Therefore the fact of the existence of an electric current is conclusive evidence of the existence of a *difference of potential* between two different points in the circuit through which the current flows.

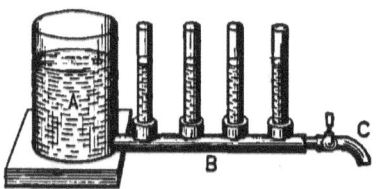

FIG. 53. Hydraulic Illustration of Electric Potential.

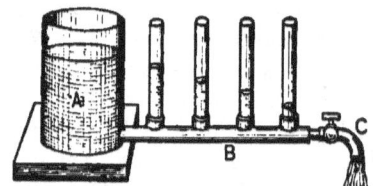

FIG. 54. Hydraulic Illustration of Uniform Fall of Potential.

---

[12] This excellent illustration is from Professor ELROY M. AVERY's *Elements of Natural Philosophy,* of which the chapter on electricity and magnetism has been separately published by Sheldon & Co., New York.

Graphic Illustration of the Electric Circuit. 71

144. Fall of Potential Proportionate to Resistance.—The fall of potential between any two points in a circuit bears the same ratio to the fall of potential in the whole circuit that the resistance between those points does to the total resistance of the circuit. In other words, *in the whole or any portion of a circuit, the fall of potential is always in proportion to the resistance.* In Fig. 55, the horizontal pipe is in two portions of different diameters, and in this case it will be observed that the fall of the pressure is more rapid along the smaller than along the larger section.

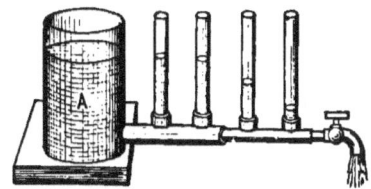

FIG. 55. Hydraulic Illustration of Varying Fall of Potential.

145. Graphic Illustration of the Electric Circuit.—We may represent by a diagram all the essential characteristics of the electric circuit in a manner first pointed out by Ohm in 1828. For example,

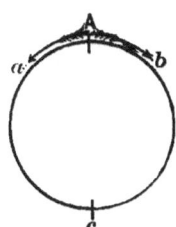

Fig. 56. Geometrical Illustration of Ohm's Law.

let the ring in Fig. 56 represent a conductor of uniform resistance having a source of electricity at the point A. The electricity from this point will be diffused over both halves of the ring; the positive going toward *a* and the negative toward *b*, both uniting at *c*. As the conductor is assumed to be homogeneous, it follows that equal quantities of electricity traverse all sections of the ring at the same time (139). If we assume that the flow of the current from one cross-section of the ring to another is due to the difference of potential which exists between the two points (143), and that the quantity which passes is proportional to this difference of potential (144), it follows that the positive and negative currents, proceeding in opposite directions from A, must exhibit a decrease in potential the farther they recede from the starting point. This decrease in potential may be graphically represented in a diagram, the analogy of which to the hydraulic apparatus of Fig. 54 will be apparent upon inspection and comparison. Suppose the ring of Fig. 56 to be stretched out in a straight line A A′, Fig. 57. Let the vertical line A B (technically termed an *ordinate*) represent the positive potential at A, and A′ B′ in like manner the nega-

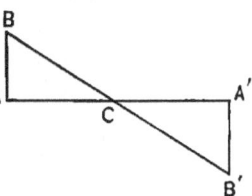

Fig. 57. Illustration of Uniform Fall of Potential.

tive potential at A'; then the line B B' will denote the value of the potential in all parts of the circuit by the correspondingly varying lengths of the vertical ordinates at any point between A c or c A'. The quantity of the current is proportional to the steepness of the fall. This may be considered also as a graphic representation of Ohm's law (127).

146. **Fall of Potential in a Non-homogeneous Circuit.**—
In practice, in the circuits employed in telegraphy, the conductor is never homogeneous, but, like the water-pipe referred to in (144), is made up of several conductors of varying conductivity. To illustrate this condition of things in a diagrammatic form, let the conductor A A', Fig. 58, consist of two portions having respectively different cross-sections. If we assume the cross-section of A d, for example, to be greater than that of d A' in the proportion of 3 to 2 ; then if equal quantities of electricity pass through all sections in equal times, as stated in (139) and (141), the difference of potential between the extremities of the thicker wire will be only two thirds what it would be in the case of the thinner wire of equal length. Hence, the fall or drop in potential will be less in the thick than in the thin wire, as shown by the line B c, in Fig. 58. The greater therefore the resistance of the conductor, the greater the fall of potential. This result is expressed in the following law:

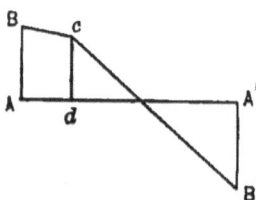

Fig. 58. Illustration of Variable Fall of Potential.

In any electric circuit, the fall of potential is directly as the specific resistances (117) *of the several conductors composing it, and inversely as the area of their cross-sections.*

The simplest circuit, therefore, when laid out in diagrammatic form, exhibits a series of gradients expressing the potential of its various parts.

147. **Electrostatic Capacity.**—A body charged with electricity in a static condition, as, for example, a long submarine cable, a condenser (317) or the well-known apparatus called the Leyden jar, is said to be in a state of *electrification*. This effect is also observable upon well-insulated land lines of considerable length, and is one which in certain special methods of telegraphy needs to be taken into consideration, as will hereafter appear. The quantity of static electricity (82) thus held by any conductor, or that which any body is capable of containing, is termed its *capacity*. This is often called also *electrostatic capacity* and *inductive capacity*.

148. **The Farad.**—The unit of capacity is called the *farad*,[13] but the capacities required to be measured in telegraphy being usually very small, they are more conveniently expressed in *micro-farads* (p. 60, note 4). Further explanation of this subject is reserved until the effects of static electricity upon telegraph lines require consideration.

149. **Power, or Rate of Work.**—It has been stated elsewhere (92) that every electric current is capable of doing a certain amount of work. This definite amount of work may, however, obviously be done in a greater or less *length of time*, that is to say, at a different *rate*, and this *rate of work* is called *power*.

150. **The Watt.**—The electric unit of power, or rate of working, is called the *watt*.[14] It equals 1 volt multiplied by 1 ampère, or $\frac{1}{746}$ of a mechanical horse-power. *In any circuit the power equals the square of the current in ampères multiplied into the resistance in ohms.*

TABLE VI.
SYNOPSIS OF PRACTICAL UNITS.[15]

Unit.	Symbol.	Name.	Derivation.	Value. C.G.S	Equivalent.
E. M. F.....	E	Volt	Ampère × Ohm	10^8	.926 standard Daniell cell.
Resistance..	R	Ohm	Volt ÷ Ampère	10^9	1.01367 B. A. Units (?).
Current.....	C	Ampère	Volt ÷ Ohm	10^{-1}	
Quantity....	Q	Coulomb	Ampère per sec.	10^{-1}	.0000105 gram of hydrogen liberated per second.
Capacity....	K	Farad.	Coulomb ÷ Volt	10^{-9}	
Power.......	P	Watt	Volt × Ampère	10^7	.0013405 or $\frac{1}{746}$ h.-power.
Work.......	W	Joule	Volt × Coulomb	10^7	.7373 foot-pounds.
Heat.......			Ampère^2 × Ohm	10^7	.238 caloric.

[13] FARADAY (MICHAEL), a distinguished chemist and natural philosopher; born in Newington, England, 1791. He received but little education, and while young was apprenticed to a bookbinder. While working at this trade, a scientific book fell into his hands, which he read with avidity, and was thus led to devote himself to the study of electricity. In 1813 he obtained the appointment of chemical assistant under Sir Humphry Davy at the Royal Institution. In 1821 he discovered magnetic rotation. In 1831 he began the publication of his *Experimental Researches in Electricity*, beginning with the induction of electric currents (151) and the evolution of electricity from magnetism (73). Three years later he discovered the principle of definite electrolytic action (154). His original papers, including a wide range of contributions to modern science, are too numerous to mention in detail. In 1833 he was appointed professor of chemistry in the Royal Institution, which chair he continued to hold until his death. He was a member of many learned societies of Europe and America. Died 1867.

[14] WATT (JAMES), an eminent mechanical engineer, born at Greenock, Scotland, 1736. Under his father he acquired a knowledge of mathematical instrument-making. When nineteen years of age he went to London, but soon returned and settled at Glasgow, where, under the patronage of the university, he subsequently immortalized himself by the invention of the steam-engine. Died 1819.

[15] MUNROE and JAMIESON'S *Pocket-Book of Electrical Rules and Tables*, p. 13*.

151. Current Induction.—An electric current traversing a conductor has a capacity of setting up or giving rise to a temporary current in a neighboring conductor. This effect is called *volta induction*, or, more commonly, *current induction;* and the temporary current thus produced is called the *induced* or *secondary current*. The originating current in such a case is termed the *primary* or *inducing current*. This effect is sometimes observed to take place between two long and well-insulated telegraph lines which are situated parallel and near together for a great distance. The flow of the secondary or induced current is in a direction contrary to that of the primary or inducing current.

152. Electrical Dimensions of the Voltaic Cell.—The practical value of any type of cell for a given purpose depends upon what is known as its *electrical dimensions*, and upon its *constancy*. The first property determines the quantity of electricity which it is capable of producing in a given time; the second property the length of time it is capable of maintaining such action.

153.—E. M. F. and Resistance of the Cell.—The electrical dimensions of a cell are stated in terms of its *e. m. f.* and its *internal resistance*. The first depends upon its chemical reaction, without reference to size, and the second is practically uninfluenced by any considerations other than the conducting power of the solutions (126), the area of their cross-section (131) and the temperature.[16]

The duration of the cell depends upon the quantity of material it contains and upon the energy of the chemical action within it. The gravity cell, described in Chapter II., has an *e. m. f.* of 1.07 to 1.08 volts, and when in good condition an average resistance of 3 to 4 ohms.

154. Quantity and Cost of Materials Consumed in the Battery.—The subjoined table shows the theoretical consumption and deposition of material in each gravity cell per ampère per hour, in fractions of an avoirdupois pound, by the aid of which the cost of producing any given current may be ascertained when the price of materials is known.[17]

---

[16] Heat increases the *e. m. f.* of a sulphate of copper cell; it does so by affecting the solubilities of the two salts and supplying externally the energy absorbed in solution. Between 32° and 52° Fahr. there is a difference of .01 volt; between 50° and 60° also .01, and between 50° and 100° about .025. J. T. SPRAGUE: *Electricity*, etc. (2d Ed.), p. 141.

[17] The electro-chemical equivalent of zinc is here taken as 0.00033696 grams per ampère per second, according to the determinations of Rayleigh and Kohlrausch. A table of electro-chemical equivalents, calculated from Rayleigh's results, is given by GEORGE B. PRESCOTT, Jr., *Electrical Engineer*, iv. 7.

Quantity and Cost of Materials Consumed.

TABLE VII.
CHEMICAL EQUIVALENTS.

Material.	Atomic weight.	Lbs. per ampère hour.
Zinc consumed...............	64.9	.0026749
Sulphate of copper consumed.	249.5	.0102810
Copper deposited............	63.0	.0025949

Experience shows that, owing in part to local action (57: 7), the actual consumption of zinc is greater than the theoretical, while the consumption of $s.\ c.$ and the deposit of copper are found to approximate quite closely to theoretical requirements. The greater part of the actual zinc-waste in practice is due to the unconsumed residue of each zinc, which finally has to be thrown out. (61a.)

155. The following examples show how this computation is made. Suppose 1 gravity cell is employed to operate a certain telegraphic instrument, whose magnetizing coil has a resistance of 3.7 ohms, and is connected with the cell by 30 feet of No. 18 copper wire. Required the theoretical cost per month of maintaining the cell, when used from 8 a.m. to 8 p.m. every day.

Average resistance of cell (assumed or measured)........3 ohms.
Resistance of 30 ft. No. 18 copper wire (table, p. 94)......0.2 "
Resistance of coil of instrument.......................3.7 "

Total resistance of circuit......................6.9 ohms.
Hence 1.07 volts (153)÷6.9 ohms=0.155 ampères.

From the table (154) we have:

Sulph. cop. consumed.. .01028   ⎫   ⎧ .5736 lbs. at 10 cts. = $0.057
Zinc consumed........ .00267   ⎬ × .155 (amp.)   ⎨ .1490 lbs. at 16 cts. =   .024
                                 ⎪ × 360 (hrs.)   ⎪   ──────
                                 ⎪   ⎪ $0.081
Cop. deposited (deduct) .00260 ⎭   ⎩ .1451 lbs. at 16 cts. =   .023
                                                                               ──────
                                                                                $0.058

156. Again, suppose two telegraph lines supplied from one battery of 100 cells, each line carrying a current of 25 milliampères (123); required the theoretical consumption per month of material, working 24 hours per day.

Sulph. cop. consumed.. .01028   ⎫   ⎧ 37.00 lb. at 10 cts. = $3.70
Zinc consumed........ .00267   ⎬ × .05 (amp.)   ⎨ 9.61 lb. at 16 cts. = 1.54
                                 ⎪ × 720 (hours)   ⎪   ──────
                                 ⎪ × 100 cells.   ⎪ $5.24
Cop. deposited (deduct) .00260 ⎭   ⎩ 9.36 lb. at 16 cts. = 1.50
                                                                              ──────
                                                                               $3.74

76 *Laws and Conditions of Electrical Action.*

We find, therefore, that the theoretical net cost of materials consumed in a battery under the conditions given, is less than 4 cts. per cell per month. In practice, it is usually from 4 to 5 cts.[18]

157. Production of Electricity in Proportion to Material Consumed.—An idea very common among amateur electricians is that it may be possible to make some change in the proportions or arrangement of the gravity battery by which its power may be increased without a corresponding expenditure of material. This is a fallacy. Electricity may in one sense be regarded as a constituent of zinc, which is set free when that metal combines with oxygen,[19] and hence the quantity of electricity evolved in a voltaic cell can never exceed a certain ratio to the weight of zinc consumed. The invariable laws of chemical combination teach us, moreover, that the consumption of $s.\ c.$ and the deposition of copper must in all cases maintain a fixed ratio to the consumption of zinc.[20]

158. Consumption of Material in a Series of Cells.—Admitting the consumption of material in each cell, when two or more cells are in series, to be in proportion to the quantity of current by which the series is traversed, it follows that the cost of material (the external resistance remaining constant) must be as the *square of the number of cells in series*, and not in the simple ratio of the number of cells. Thus if we increase the number of cells threefold, we have three times as many cells and three times the quantity of current traversing each cell, so that the consumption of material will necessarily be ninefold.

159. Electrical Dimensions of the Edison-Lalande Cell.—This element has a comparatively low $e.\ m.\ f.$ (0.70 to 0.75 volts), but on the other hand its internal resistance is very small, and its local action almost inappreciable. Such a cell is well suited for telegraphic work. The diagram Fig. 59, exhibits the results of a test of 4 large cells, maintained in action in series with an external resistance of 0.8 ohms continuously for 108 hours. Such a current would suffice to supply 10 or 12 telegraph lines at the same time.

160. Effect of Temperature upon the Resistance of Metallic Conductors.—It has been stated (118) that the resistance of all conductors is affected by temperature. Unless otherwise specified, the resistance of electrical conductors is customarily assumed to be taken at 60° Fahr.

---

[18] L. BRADLEY in *The Telegrapher*, iii, 53; F. L. POPE, in *the same*, vii, 345; *Scientific American* (n. s.) ix, 184.
[19] This suggest'on is due to Latimer Clark. See his *Electrical Measurement, p.* 168.
[20] See note 2, p. 12; also (154).

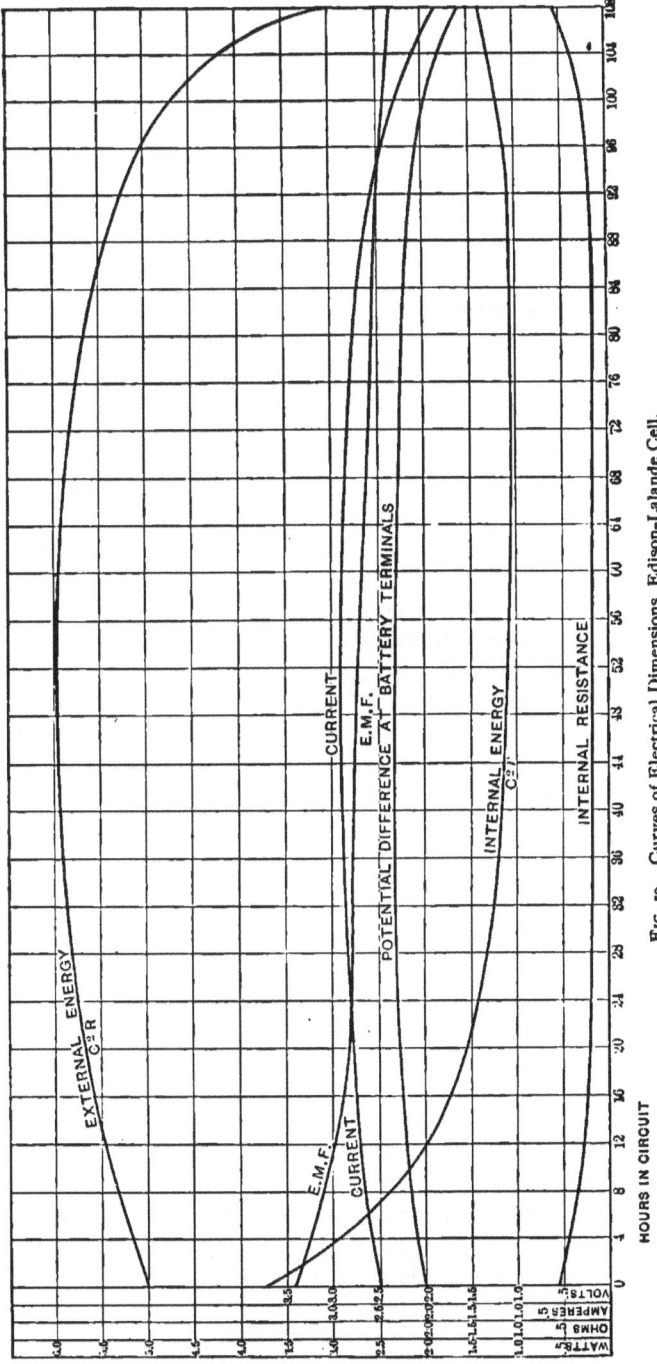

Fig. 59. Curves of Electrical Dimensions, Edison-Lalande Cell.

161. According to Müller, the percentage of increase in resistance of some of the metals most employed in telegraphy between $0°$ and $70°$ Fahr. is as follows:

Iron	8 per cent.
Copper	6.1 per cent.
Platinum	6 per cent.

The difference in the measured resistance of a telegraph line of iron wire may therefore vary as much as 13 per cent. between the extremes of summer and winter temperature in the northern portions of the United States. The resistance of German-silver and platinum-silver alloys vary but little with temperature, and hence standard resistances are made from wires of these artificial metals.

162. **Effect of Temperature upon Resistance of Liquids.** —The liquid mass which acts as a conductor in a voltaic cell undergoes considerable variation in resistance with changes of temperature. Of the different voltaic combinations in general use, the sulphate of copper cell is most affected in this way. Hence in experiments with this cell, it is important that the temperature be kept constant, or that frequent measurements should be made of the internal resistance and allowance made therefor.

163. **Effect of Temperature upon Resistance of Daniell Cell.**—Three series of tests of the Daniell sulphate of copper cell were made by Preece; in the two first cases, the *s. c.* solution was saturated at all temperatures, while the *s. z.* solution had the same density throughout the period of observation, being saturated at about $57°$ Fahr. In the third case both solutions remained saturated at about $50°$ Fahr. The results are given in the diagram, Fig. 59*a*. The curve ABCDE corresponds to the case in which the *s. c.* solution was saturated at all temperatures, while the *s. z.* solution was of constant density. The curve *abcd* corresponds to the case in which both solutions remained unaltered in density. The direction of the arrows indicates the order of the experiments. In the curve ABCDE, the portion AB represents the result obtained by heating the cell from about $52°$ to $211°$ Fahr. (near the boiling point of water), and the portion BC that obtained while the same cell was being cooled from $211°$ to $35°$ Fahr., nearly the freezing point of water. A similar explanation applies to the curve *abcde*.

164. These curves clearly show:

(*a*) That when the temperature of a Daniell cell is raised from the freezing to the boiling point of water, the internal resistance of the cell decreases, abruptly at first, but more gradually afterward, falling from 2.12 to .66 ohms, or more than one-third.

Effect of Temperature upon Battery Resistance. 79

(*b*) That when a cell which has been thus heated is cooled, the resistance increases at a more rapid rate than it fell off while being heated; in other words, the resistance of a Daniell cell, within the range of temperature experimented upon, is smaller before it has been heated to a high temperature than afterward, provided the heating and cooling be not done too slowly.

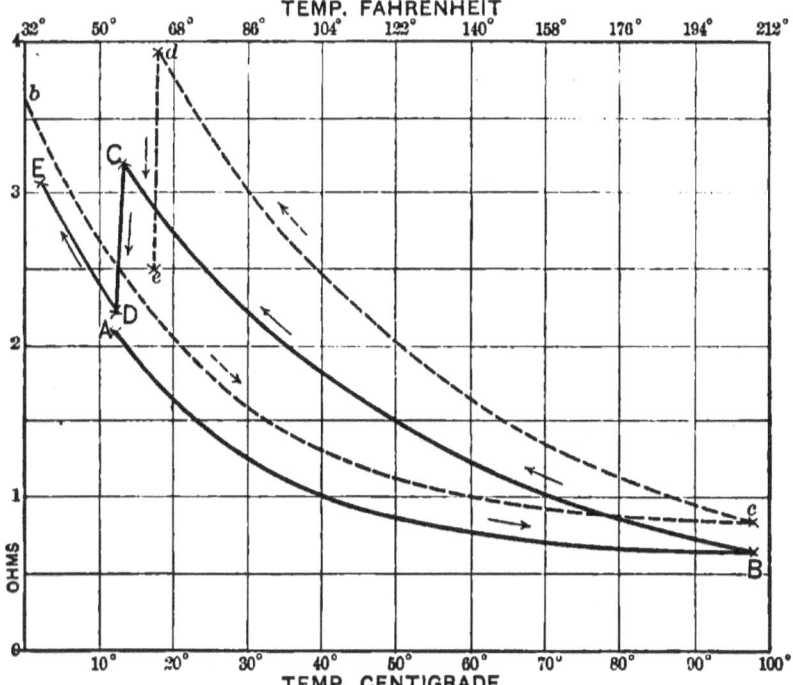

FIG. 59 a. Effect of Temperature upon Resistance of Voltaic Cell.—PREECE.

(*c*) That if the cell thus cooled down be left undisturbed at a given temperature, the resistance of the cell slowly diminishes until at last, at the end of a certain period (40 to 50 hours), it returns to the value which it had before having been heated.

(*d*) That the resistance of a Daniell cell is considerably less when the s. c. solution is more dense than when it is less dense, at any temperature.[21]

[21] W. H. PREECE: *Effect of Temperature on the e. m. f. and Resistance of Batteries.* Proc. Royal Soc. 1883; Lond. *Electrician.* x. 367.

CHAPTER VI.

THE LAWS OF ELECTRO-MAGNETISM.

165. The *Electro-Magnet*, as improved by Henry,[1] forms the most essential part of every telegraphic receiving instrument, and is the instrumentality by means of which the energy of the electric current is transformed into mechanical power, and is made to produce physical effects appreciable by the senses.

Nearly every fact of importance in connection with the phenomena of electro-magnetism has been known to experimenters and observers for half a century, but the apparently anomalous and contradic-

[1] HENRY (JOSEPH), LL.D., born Albany, N. Y., 1799; educated in the common schools of that city and in the Albany Academy, in which he became professor of mathematics (1826), and almost immediately entered upon a course of experimental investigation, during which he made numerous and important discoveries in electricity and magnetism. Although at this date the electro-magnet had become in a certain sense known, from the researches of Sturgeon (*Transactions Soc. Arts*, xliii. ; Nov. 1825), it was but a philosophic toy, in which a feeble magnetic excitation was produced by currents of small *e. m. f.* in a short circuit. Henry's first success was the invention of the *electro-magnet* as we now know it, a horse-shoe of soft iron surrounded by many turns of insulated copper wire arranged in concentric layers (186), a construction which no subsequent invention has essentially modified. He next demonstrated that the difficulty of exciting magnetic energy at a distance by an electric current, which had led Barlow in 1824 to pronounce the idea of an electric telegraph "chimerical," may be completely overcome by the use of a battery of a sufficient number of cells, arranged in series (107), provided the electro-magnet be provided with a helix having a sufficient number of turns. It was the invention of Henry's electro-magnet which first made the electric telegraph a commercial possibility, and it is worthy of note that in an article published in 1831 (*Amer. Jour. Science*, xix. 400), he pointed out the applicability of the long-coil magnet to this purpose. During the same year he constructed an apparatus for giving signals at a distance, which was operated through more than a mile of wire carried around the walls of a room in the Albany Academy. This apparatus embodied all the essential principles of the practical telegraph of to-day. The signals were produced by the polarized armature of an electro-magnet which was made to vibrate by reversal of the current (201) and to strike a bell. In 1832 he discovered the induction of a current in a coiled conductor upon itself (196). In 1832 he was elected professor of natural philosophy in the College of New Jersey, at Princeton, and in 1846 first secretary of the Smithsonian Institution in Washington, which honorable position he continued to hold until his death in 1878. His collected scientific papers have been published in 2 vols., Washington (1886). For many particulars of interest respecting the contributions of Henry to the invention of the electric telegraph, see *Life and Work of Joseph Henry*, by F. L. POPE, and "The American Inventors of the Telegraph," by the same, in the *Century Magazine*, xxxv. 924 (April, 1888). It has recently become known that he was the first to discover the phenomenon of magneto-electricity (62). See papers by MARY A. HENRY, *N. Y. Elect. Engineer*, xiii. 27 *et seq*.

Elements of the Electro-Magnet. 81

tory character of many of the results obtained has been very puzzling to the student. It was not until after the conception of the existence of a magnetic circuit (178), analogous in many of its properties to that of the electric circuit, originally due to Joule,[2] had been definitely formulated in 1873 by Rowland,[3] and its truth confirmed by the subsequent researches of Bonsanquet[4] and others, that it became possible to suggest an adequate explanation for many of the singular and apparently unaccountable facts which had been noticed by investigators.

166. **Elements of the Electro-Magnet.**—The electro-magnet may conveniently be regarded as comprising three distinct elements, the laws of each of which must be separately studied, although they all enter into the general result. These elements are (1) the wire, (2) the iron, and (3) the current.

167. It has been stated (85, *d*) that when a piece of soft iron is spirally encircled by a conductor, it is rendered *magnetic* by the passage of a current through this conductor. Such an organization constitutes an *electro-magnet* in its elementary form.

168. **Polarity of Electro-Magnet Determined by Direction of Current.**—The position of the respective poles of an electro-magnet is in all cases determined by the direction of the magnetizing current. It is usual to coil the conducting wire, coated or insulated, with non conducting material, into what is termed a right-handed helix, shown diagrammatically in Fig. 60, in which the conventional direction of the current (31) is indicated by the arrows, while the respective north and south poles induced thereby are designated by the letters N and S. Thus, if the current flows around the iron in the direction of the hands of a watch, the north pole will be at the distant end of the iron. If the current be made to flow in the opposite direction, as in Fig. 61, the polarity

Fig. 60. Electro-Magnet with Right-handed Helix.

Fig. 61. Electro-Magnet with Left-handed Helix.

[2] *Sturgeon's Annals*, iv. 58.
[3] H. A. ROWLAND: *On the Magnetic Permeability of Iron*, Phil. Mag. (4th series), xlvi. 140; also in *the same*, 1, 257, 348.
[4] R. H. M. BONSANQUET: *On Magneto-Motive Force*, Phil. Mag. (5th series), xv. 205; the same, *On Electro-Magnets*, Electrician (ond.), xiv. 291, 351.

of the iron is reversed, the north pole now being at the end where the south pole was before, and *vice versa*.

169. **Lines of Force as a Measure of the Magnetic Field.**—It has been explained (93) that a conductor conveying an electric current is surrounded by a field of magnetic force, and that in such a field, the lines of force are concentric with the conductor. These lines of force may be regarded as units, in terms of which magnetism may be expressed and measured.

The *direction* and *polarity* of the magnetic force is indicated by the direction and polarity of the lines, the total number of its lines is a measure of the *total quantity of magnetism*, while the number of them contained in a given unit of area, measured in a direction perpendicular to their direction, is a measure of the *intensity of magnetism* at that point.

> This conception of magnetic force may, perhaps, be better understood if compared to the force of gravity similarly represented. Imagine a heavy body suspended in the air, and suppose every cubic inch of the material of which the body is composed to weigh one pound. If an imaginary line be drawn to the earth from the center of gravity of each cubic inch of the suspended body, the direction of these lines would represent the direction of the force of gravity; their total number would represent the total force in pounds; while their density, or the number of lines per square inch area (measured perpendicularly to their direction), would represent the intensity of the force at that point. In precisely the same way as these lines represent the direction, amount, and intensity of the force of gravity in that body, so do the lines of magnetic force represent the direction, amount, and intensity of magnetism, except that in the latter there is no constant direction of action such as the downward force of gravity, the lines of force acting in both directions, as if trying to shorten their circuit, like a stretched rubber ring. The lines do not exist as such, any more than they do in the analogy of the force of gravity; it is merely a convenient way of representing magnetism in order to facilitate the conception and computation of problems.— CARL HERING: *Principles of Dynamo-Electric Machines*, 18.

170. An accurate knowledge of the characteristics of magnetism is of great importance in the designing and construction of dynamo-electric machinery, and it fortunately happens that recent researches in connection with this class of work have greatly enlarged our practical knowledge of the laws and conditions of magnetic and electro-magnetic action as applied to telegraphic and other apparatus of like character.

Provided we are able to calculate the intensity of the magnetic field which is produced by the influence of a known current, we have the means of calculating also the intensity of magnetism in an iron core placed within that field. When, however, the magnetization approaches the limit of intensity which the soft iron is capable of receiving, the actual magnetization always falls short of the theoretical magnetization as calculated by this rule.

171. **Unit of Magnetism.**—It is customary to express *intensity of magnetism*, or *magnetic density*, as it is sometimes termed, by the *number of lines of force per unit of cross sectional area*, measured perpendicularly to their direction. The *unit of magnetism* is the equivalent of a single one of these lines of force, and is that quantity of magnetism which passes through one square centimetre of the cross-section of a magnetic field whose intensity is unity. It has been proposed to call the magnetic unit the *gauss*.[5] To illustrate, suppose a circular loop of wire like that shown in Fig. 33, p. 41, having a diameter of 10 centimetres (3.9 in.), to be traversed by a current of 7.958 ampères. The quantity of magnetism passing through an area of 1 sq. cm. at the center of the loop will be 1 unit.[6] A magnetic field in which the number of parallel lines of force per unit area is the same in every part is termed a *field of uniform intensity*, or briefly, a *uniform field*. A good illustration of such a field is that of the earth referred to in (94). The field inclosed within a circular loop of wire like Fig. 33 is not uniform, but varies in different parts, being most intense near the circumference and least in the center.

172. **Magneto-Motive Force.**—Recurring again to the electric conductor surrounded by concentric lines of force, as shown in Fig. 32, p. 39, it is nòt difficult to understand that if we coil such a conduc-

[5] GAUSS (KARL FRIEDRICH), born in Brunswick, Germany, April 30, 1777. When very young was distinguished for his mathematical attainments ; became Professor of Astronomy and Director of the Observatory in Göttingen, 1807 ; was made, in 1816, Court Councilor and in 1845 a Privy Councilor of Hanover ; after 1821 made important improvements in geodetic methods and instruments ; and after 1831 devoted much attention to the study of terrestrial magnetism. In 1833, with the assistance of his coadjutor, WILHELM EDUARD WEBER, Professor of Physics in the University of Göttingen, he constructed an electric telegraph more than a mile in length, extending from the Physical Cabinet to the Observatory in that city. This telegraph was remarkable as being the first in which magneto-electricity (73) was used ; for the ingenious but simple method employed of using a ray of light as an index of the movement of the galvanometer needle (a plan long afterward adopted by Sir William Thomson in his well-known mirror galvanometer) ; and last, though not least, as having had an actual existence for several years ; for although at first intended for scientific purposes only, it soon came to be employed as a means of ordinary correspondence as well. (SABINE: *The Electric Telegraph*, p. 27.) Gauss died at Göttingen, 1855.

[6] What is known among manufacturers of electrical machinery as the *English unit* of magnetic induction was proposed by GISBERT KAPP (*Jour. Tel. Eng.*, xv. 518). The unit line of force adopted is equal to 6,000 *c. g. s.* lines, the sectional area of the iron being taken in square inches. The English unit, therefore, is one of these assumed lines per square inch, and is commonly termed a *Kapp line.*

1 Kapp line per sq. in. = 930 *c. g. s.* lines per sq. cm.
1 English unit = 930 *c. g. s.* units of magnetic induction.

tor into an elongated helix or spiral, technically termed a *solenoid*, Fig. 62, and cause the current to traverse it in the direction indicated by the arrows, the lines of force inclosed within the helix, being the resultant of those of the separate turns, assume the form represented

FIG. 62. Direction of Lines of Force within a Solenoid.

in the figure. In the drawing, for convenience of illustration, only a part of each line of force is shown, but it must be borne in mind that every line is in fact endless, forming a complete magnetic closed circuit returning into itself, so that different lines can never under any circumstances intersect each other. The value of the current in ampères being known, a corresponding field of definite intensity is set up within the helix. The intensity, or, as Bonsanquet calls it, the *magneto-motive force* of the field, may be readily calculated by the following:

RULE.—Multiply the number of turns in the helix by the current in ampères and divide this product (ampère-turns) by the length of the helix in centimetres; multiply the quotient by 1.2566, and the product will be the intensity expressed in lines of force per square centimetre, or if the length be taken in inches, the multiplier 0.3132 will give the quotient in lines per square inch.

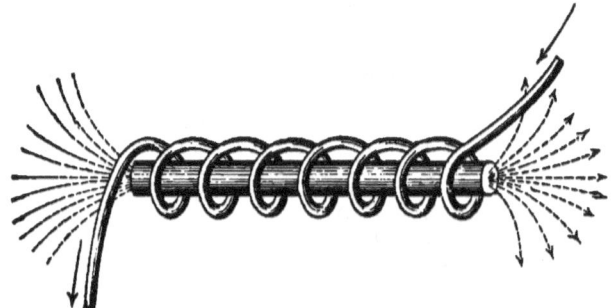

FIG. 63. Lines of Force traversing Iron Bar within the Solenoid.

173. Effect of Iron in the Helix.—If now a soft iron core be placed within the same helix, as shown in Fig. 63, the intensity of the field is materially increased, or in other words the number of lines of force per unit of cross-sectional area is greatly augmented. The strength of field due to the presence of the coil and its contained iron is termed *magnetic induction*. The difference between the number of lines per unit of area, with and without the iron, evidently gives the value of the magneto-motive force due to the iron alone. This difference may be stated roughly as about 100 to 1 for soft iron of average good quality.

174. Effect of Magnetization upon Soft Iron. — The graphic diagram, Fig. 64, was plotted from a series of observations made with a *magnetometer*[7] upon a rod of unannealed iron 10 cm (3.9 in.) long and 4.3 mm. (0.169 in) in diameter, placed within a helix of 135 turns.[8] The values of the current in ampère turns are plotted out upon the horizontal, and those of the magnetic forces upon the vertical scale. The resulting curve takes the form shown in the figure by the line o B. It will be seen to consist of two parts; one part which rises at a more or less steep angle, and which for some distance from its origin at o continues nearly straight to the point 1, and another part B 2, also nearly straight, but which is inclined at a much less angle to the horizontal, these two parts being joined by

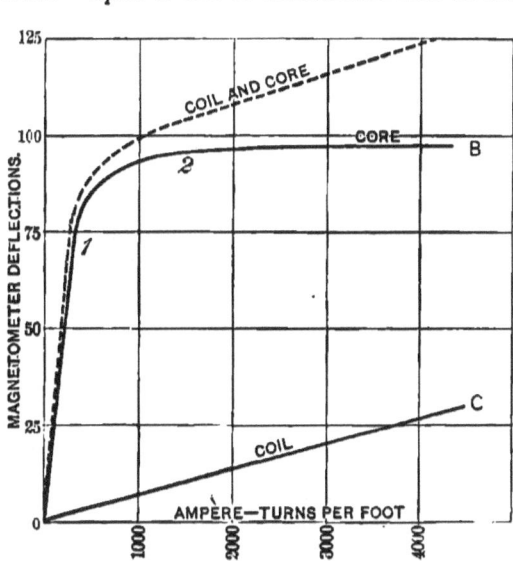

FIG. 64. Relation of Current to Magnetic Force.

---

[7] The magnetometer is an instrument for the measurement and comparison of magnetic forces. It consists essentially of a magnet or needle delicately suspended in the magnetic meridian, and provided with a pointer or index, usually in the form of a ray of light reflected by a small mirror. By placing the stationary magnet whose force is to be measured at a measured distance east or west of the suspended needle, with one of its poles pointing directly toward it, it is easy, by observing the angle of deflection of the needle, to measure the attractive force of the magnet pole. For a simple apparatus and method of performing this operation see J. TROWBRIDGE: *New Physics*, 131.

[8] KENNELLY and WILKINSON: *Practical Notes for Electric Students*, 220.

a curved portion 1, 2. The first-mentioned part of the curve corresponds to the state of things when the iron core is unsaturated; the latter part to the state when the core is more than half saturated; while the curved intermediate portion corresponds to the intermediate state during which the core is approaching saturation (177). In the curve of results of an electro-magnet two effects are in reality combined; that of the magnetism of the iron core, and that of the magnetic action of the coils through which the current is flowing; this joint effect is shown in the dotted line. It is easy to separate these two values, for if the iron core be removed, and the magnetic effect of the coils alone be observed, a new set of data are obtained which, when plotted out, will yield the more gently sloping line o C. From this line two conclusions may be drawn: it slopes at a small angle, because (1) the magnetic effect of the coils is small compared with that of the iron core. It is quite straight, because (2) the magnetic effect of a coil (which of course is not capable of saturation) is exactly proportional to the strength of the current by which it is traversed, throughout the entire range of the experiment.

175. The following series of determinations, made with a coil of 500 turns surrounding an iron core 10 cm. (4 in.) long and 1 cm. (13·32 in.) in diameter, further illustrate this matter. The figures in the last column are the values of the *magnetic moment*[9] as calculated from the deflections produced in a magnetometer.[10]

Ampères.	Ampère-turns.	Magnetic Moment.
0.00	0	128
0.22	110	1224
0.39	195	1920
0.98	490	4608
1.33	665	5924
3.65	1825	17472
4.6	2300	21088
9.2	4600	27875
9.4	4700	28750

The above values, when plotted out, will give curves similar in form to those shown in Fig. 64. The values of magnetic moment

---

[9] The *magnetic moment* of a magnet in c. g. s. measure is the product of the strength of its magnetic pole in dynes (191) multiplied by the distance between its poles in centimetres. The *intensity of magnetization* of a magnet is the ratio of its magnetic moment to its volume.

[10] SILVANUS P. THOMPSON: *Dynamo-Electric Machinery*, (2nd edition), 355.

for telegraphic magnets, when plotted out in the same way, will fall upon the lower half of the steep, straight portion of such a curve.[11]

176. Magnetic Saturation.—It will be seen, therefore, that the proportion of ampère-turns to magnetic intensity, referred to in (174), holds good only through a certain range of magnetic increase. When the intensity has reached a certain point, the iron becomes, from that point onward, less and less susceptible to further magnetization, and though, strictly speaking, the point of absolute saturation can never be reached, there is a practical limit which cannot be exceeded.[12] The approach of saturation is well exhibited in the core curve in Fig. 64, which begins to deflect when the magnetizing force reaches the vicinity of 500 ampère-turns.

The cores of the electro-magnets of modern telegraphic apparatus seldom exceed 0.5 in. in diameter. It has been experimentally proved that the approach of saturation in a core of this dimension is not reached with less than about 500 ampère-turns, which is some 3 times the degree of magnetization ordinarily employed in telegraph magnets used in local circuits, while that employed in magnets used in main circuits is still less.

177. Magnetization Proportional to Ampere-turns.—An important principle in electro-magnetism is, that precisely the same magnetic effect may be obtained from a few turns of wire and a large volume of current as from a great number of turns and a small current, provided only that the number of ampère-turns remains the same. This necessarily follows from the fact that the same amount of work is done in the wire by the circuit in each case (92).

178. The Magnetic Circuit. —In the practical application of the electro-magnet for telegraphic and other like uses, it is not usual to make it in the form of a straight bar. Much better results are attained by bending the bar into the form of a **U**, or "horseshoe," as shown in Fig. 65, which enables an armature to be applied to it in such a manner as to form a complete

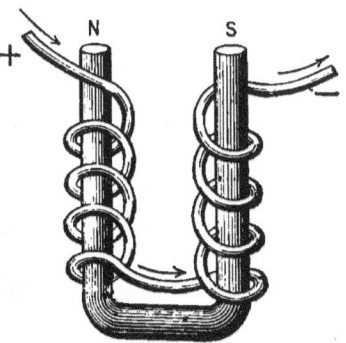

FIG. 65. Principle of Horseshoe Electro-Magnet.

---

[11] For an experimental investigation of the relation between the diameter of the core, the total magnetizing force of the coil, and the force of attraction, see paper by E. L. FRENCH : *Electrician and Elect. Eng.*, v. 445.

[12] The limit of magnetization in good wrought-iron is about 125,000 (*c. g. s.*) magnetic lines per sq. in., or 20,000 per sq. cm.—S. P. THOMPSON : *The Electromagnet*, 82, 83; ibid., *Dynamo-Electric Machinery* (4th Ed.), 148, 149.

The Laws of Electro-Magnetism.

magnetic circuit. Inasmuch as magnetism is now known to be a phenomenon pertaining to the internal molecular structure of iron, the preferable method of treating the subject is to look upon that metal as a substance which is a good conductor of the magnetic lines of force, or, as it is expressed in modern scientific language, possessing a high degree of *magnetic permeability*.[13]

179. **Magnetic Permeability.**—This characteristic may be best defined as a *numerical co-efficient* which expresses the ratio between the number of magnetic lines formed in a space containing nothing but air, as in Fig. 62, and as denoted by the value of the line o B in Fig. 64, and the number formed in a space filled with a given quality of iron, as in Fig. 63, and as denoted by the value of the dotted line in Fig. 64.[14] This ratio differs for different qualities of iron, and hence we say that the permeability of the iron differs accordingly.

The higher the co-efficient of permeability, the less, so to speak, is the *magnetic resistance*, and the more suitable is the iron for the purposes of an electro-magnet. On the other hand, the permeability of air and of most substances other than iron is comparatively very small.

180. **Law of the Magnetic Circuit.**—In (172) the method of calculating the magneto-motive force of a magnetic circuit has been given. We have next to find the *resistance* which the magnetic circuit offers to the passage of the lines of force, a property which has appropriately been termed by Dr. O. J. Lodge *magnetic reluctance*. The total magnetism of the circuit, called the *magnetic flux*, will be the quotient of the magneto-motive force divided by the reluctance. The similarity of the law of the magnetic circuit to the law of the electric circuit, heretofore referred to as Ohm's law (127), will be apparent upon inspection.

181. **Determination of Magnetic Reluctance.**—If the magnetic circuit is a simple closed ring of iron, the magnetic reluctance may be calculated precisely in the same manner that we calculate the resistance of an electric circuit. The value of the reluctance is directly in proportion to the length of the iron, inversely as the area

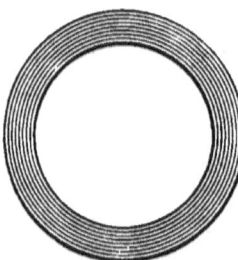

Fig. 66. Lines of Force in Endless Iron Ring.

[13] FARADAY: *Exper. Res.*, iii. 426; Sir W. THOMSON: *Papers on Electricity and Magnetism*, 484.
[14] ROWLAND: *Phil. Mag.* (5th series), xlvi. 140.

of its cross-section, and is also inversely proportional to its permeability. But if, instead of a homogeneous ring of iron, the circuit be made up of different parts, differing in their magnetic reluctance, it becomes necessary to determine the reluctance of each part separately, and then add them together, as in the case of an electric circuit of like character (118). For example, Fig. 66 shows the lines of force in an endless iron ring. Fig. 67 is a similar ring cut in two, leaving an air-gap between the severed ends. It has been stated that the permeability of air is far less than that of iron (179). The reluctance of the air-gaps to the magnetic lines may be taken roughly at 100 times that of a mass of soft iron of good quality of the same form and dimensions. The case of the divided ring of Fig. 67 is equivalent to that of the horseshoe magnet and its armature shown in Fig. 65, when the armature is a little way removed from the poles, and is the condition which is constantly met within the operation of ordinary telegraphic apparatus. The lines of force traverse the armature in passing from one pole to the other.

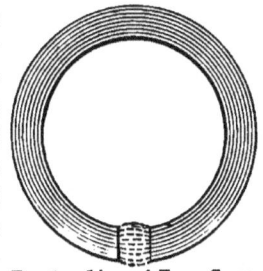

FIG. 67. Lines of Force Crossing Air-gap in Magnetized Ring.

182. Ratio of Attractive Force to Distance.—It is stated in many text-books that the attractive force exerted by an electromagnet upon its armature varies inversely as the square of the distance between them. This proposition, known as Coulomb's law, would be true, if it were true that the magnetic forces are concentrated at a focal point in each pole, and that this disposition of it remains unchanged by the movement of the parts in response to the magnetic attraction. But in fact there is not, and from the nature of the case cannot be, any one law which correctly expresses this relation under all conditions. It necessarily differs with every alteration in the form of magnet and armature, and with every change in their positions with reference to each other. This is well shown in experiments[15] made with an electro-magnet having a core formed from a round bar 19 in. long and 1 in. thick, bent into a horseshoe, with its poles 1.25 in. apart. The distance of the armature from the poles was determined by the interposition of sheets of rolled brass .00416 in. thick, the required number of these sheets for each experiment being strongly pressed together and soldered at the edges. The following table gives the results in weights lifted,

[16] DANIEL DAVIS, Jr.: *Manual of Magnetism* (12th Ed., 1857), 152.

with various thicknesses of brass sheets, numbered from 1 to 10, interposed between the magnet and the armature:

Distance.	Weight Lifted, Grains.	Product.
0	82,000	......
1	35,000	35,000
2	25,000	50,000
3	20,000	60,000
4	15,500	62,000
5	12,100	60,500
6	11,300	67,800
7	9,300	65,100
8	7,400	59,200
9	6,500	58,500
10	5,500	55,000

The corresponding curve is plotted in Fig. 68. The rapid increase in the attractive force as the armature approaches the poles of the magnet is shown in a striking manner.

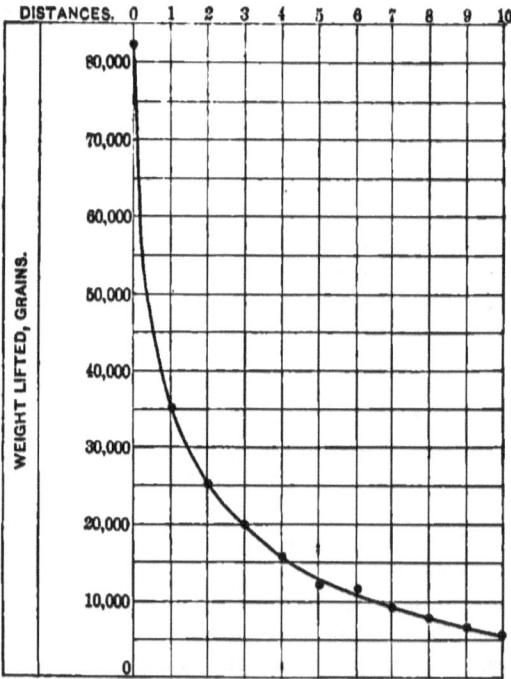

FIG. 68. Ratio of Decrease of Magnetic Attraction to Distance.

Construction of Telegraph Magnets. 91

183. Construction of Telegraph Magnets.—Fig. 69 is a representation of an electro-magnet, such as is usually employed in telegraphy. The drawing is the actual size and proportions of a type of magnet largely used by some of the most successful American instrument-makers. The iron portion of the magnet, of the best

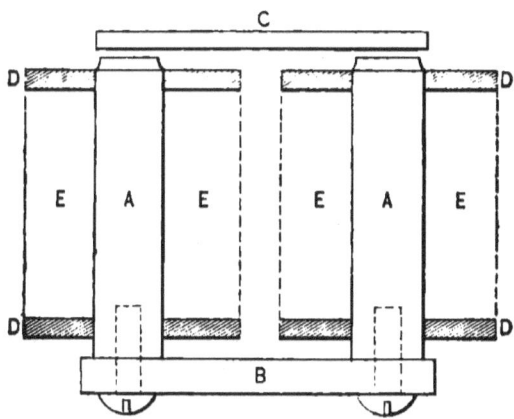

FIG. 69. Telegraphic Electro-Magnet—Full Size.

Swedish, Norwegian, or Lowmoor soft iron, consists of the following parts: (1) the *core* proper, which is cylindrical in form, and is the part around which the wire is coiled: it is made in two parts, A A, usually termed the legs or branches of the magnet; (2) a rectangular bar, B, which serves to unite the two parts of the core (which are secured to it by screws), and is termed the *yoke;* and (3) the *armature* C, which, as has been shown, is really part of the magnet, being the movable portion by means of which the magnetic force is exerted.

184. Theoretical Proportions of Telegraph Magnet.— The best theoretical proportions to secure the *maximum magnetic effect* from a given quantity of current, has been found to be to make the four parts of equal length, the yoke being of somewhat greater cross-section than the cores, and the armature of equal cross-section, but broader and thinner than the yoke. But inasmuch as quickness of movement is one of the most important considerations in telegraphic apparatus, experience has demonstrated that these theoretical proportions may be modified with practical advantage.

The dimensions and proportions of the iron cores of electro-magnets have been the subject of numerous experiments in order to determine the most favorable conditions in respect to the two qualities essential in telegraphic instruments: (1) maximum attractive force with a given current, and (2) quickness of action. These properties are in their nature antagonistic, and hence it is necessary in practice to sacrifice to a certain extent the first-named desideratum in order to more completely secure the second. The results of the investigations referred to have shown that the outer diameter of the coils or helices ought to be three times that of the cylindrical cores, and that the length of each coil or helix should be equal to its diameter. These proportions are exemplified in Fig. 69, and approximate closely to those most commonly used at the present day in the United States. The magnetic intensity developed in the iron, within certain limits elsewhere set forth (174), being proportional to the quantity of current traversing the wire (measured in ampères), and also to the number of convolutions or turns of the wire, we may express the magnetism developed in the iron as a certain number of *ampère-turns*.

185. **Effect of Position of Windings.**—It makes no appreciable difference upon what portion of the core any particular turn is wound, nor does the fact that some of the turns may be close to the iron and others at a greater distance from it, appreciably modify the result, within the limits of the dimensions of the magnets used in telegraphy.

186. **The Helix or Coil.**—Upon the cylindrical cores of the magnet are fixed flanges or collars D D (Fig. 69), of hard rubber or other like material, which, in connection with the cores, form spools or bobbins upon which the magnetizing coil is wound in superposed concentric layers. The space E E, which is designed to contain the wire, has its boundary indicated by a dotted line.

187. **Relation of Thickness and Length of Wire to Number of Turns.**—The length of any given wire which can be

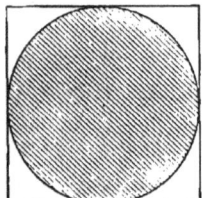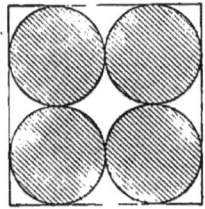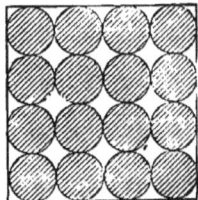

FIG. 70. Illustration of the Law of Diametrical Squares.

wound within a space of given dimensions, such as the space E E, Fig. 69, is inversely in proportion to the square of the diameter of the wire. This will appear from the diagram Fig. 70, in which are shown within a space 1 in. square, the outlines of 1 wire 1 in. diame-

Relation of Thickness and Length of Wire. 93

ter, 4 wires ½ in. diameter, and 16 wires ¼ in. diameter, all of which occupy precisely the same area of cross-section in the spool. The number of turns which can be put within a given space is also inversely as the square of the diameter of the wire, measured to include its insulating covering.[16] As the electrical resistance of a wire is directly as its length and inversely as its sectional area or the square of its diameter (118), it will be obvious that the number of turns in the coil of any electro-magnet must have a direct and invariable relation to its resistance, and hence *the resistance of a coil may be taken as a measure of the number of turns of wire it contains.* This is convenient in practice, inasmuch as the resistance is easily determined by proper apparatus, while it is not so easy to find the number of turns in a coil after it has been wound. It is for this reason, and not because the resistance in itself has anything to do with the matter, that it has become customary among telegraphists to classify electro-magnets by reference to their measured resistances.

It is difficult to wind a magnet coil neatly and accurately without the aid of machinery. It may be done in a common lathe, but amateurs generally will find it more convenient to use one of the little machines now made for the purpose, such as that shown in Fig. 71. The hub, which is seen lying on the table, is screwed into the end of the cylindrical core upon which the coil is to be wound, and its other end screws on the spindle at the top of the machine. The operation of winding is sufficiently explained by the illustration.[17]

FIG. 71. Machine for Winding Magnet-Helices.

188. Dimensions and Resistances of Magnet Wires.— The following table gives the properties of the different sizes of cop-

[16] A very convenient rule for calculating the windings of the coils of two different electro-magnets of the same type, but of different dimensions, is given by Sir William Thomson, and is as follows: Similar iron cores similarly wound with lengths of wire proportional to the squares of their linear dimensions will, when excited by equal currents, produce equal intensities of magnetic field at points similarly situated in with respect to them. Professor Silvanus Thompson has also pointed out as a corollary that similar electro-magnets of different dimensions must have ampère-turns proportional to their linear dimensions, if they are to be magnetized up to an equal degree of saturation.

[17] These machines are made by H. Anderson, Peekskill, N. Y.

per wire most used for the helices of galvanometers and telegraphic magnets. It is taken from one calculated by George B. Prescott, Jr., on the basis of Dr. Matthiessen's standard, viz.:

1 mile of pure copper wire of $\frac{1}{16}$ in. diam.=13.59 ohms at 59.9° Fahr.[18]

TABLE VIII.

DIMENSIONS AND PROPERTIES OF COPPER MAGNET WIRES.

American Gauge No.	Diameter Mils.		Area.		Weight and Length.		Resistance at 75° Fahr.		
	Bare Wire.	Silk-covered Wire.	Circular Mils (d^2) 1 Mil = .001 in.	Square in. $d^2 \times .7854$	Lbs. per 1000 Feet.	Feet per Lb.	Ohms per 1000 Feet.	Feet per Ohm.	Ohms per Lb.
18	40.3	42.6	1624.3	1275.7	4.91	203.8	6.39	156.47	1.30
20	32.0	34.0	1021.5	802.3	3.09	324.0	10.16	98.401	3.29
22	25.3	27.3	642.7	504.8	1.94	515.1	16.15	61.911	8.32
24	20.1	22.2	404.0	317.3	1.22	819.2	25.69	38.918	21.05
26	15.9	17.9	254.0	199.5	.77	1302.6	40.87	24.469	35.23
28	12.6	14.2	159.8	125.5	.48	2071.2	64.97	15.393	134.56
30	10.0	11.6	100.5	78.9	.30	3294.0	103.30	9.681	340.25
32	8.0	9.0	63.2	49.6	.19	5236.6	164.26	6.088	860.33
34	6.3	7.3	39.7	31.2	.12	8328.3	261.23	3.828	2175.50
36	5.0	6.0	25.0	19.6	.08	13238.8	415.24	2.408	5497.40

The thickness of silk-covered wire is approximate only; it varies somewhat with different makers.

The figures in the table refer to a single covering of silk. For a double-covered wire, add the difference between the figures in the second and third columns to the figures in the third column.

189. **Thickness of Spaces between Turns of Wire.**—The thickness of a covered wire or of its covering cannot be correctly determined by the process of direct measurement by a gauge (192), though it may be approximated by the careful use of such a micrometer caliper as that shown in Fig. 73. The most accurate method is to measure the longitudinal space occupied by a number of turns when closely wound upon a mandril or small cylinder; divide this length by the number of turns, and from the quotient subtract the diameter of the copper wire measured by the micrometer caliper, and divide the result by 2, which will give the thickness of the covering.[19]

[18] *Electrician and Elec. Eng.*, iv. 217.

[19] Helices made of bare copper wire, accurately wound by machinery in such a manner as to leave an air-space of 1 mil. (.001 in.) between each two adjacent turns, and having the successive layers separated by thin paper, have been much used in the United States with very satisfactory results.

Instruments for Gauging Wire. 95

190. American Standard Wire Gauge.—Great confusion formerly existed, both in this and other countries, in respect to wire gauges, designated as the custom is by progressive numbers, there having been almost as many so-called standards as there were different manufacturers. The Brown & Sharpe Manufacturing Co., of Providence, R. I., some years since established a gauge in which the actual thickness of wires designated by successive numbers is made to diminish in a true geometrical progression. Under the name of the *American gauge*, this has now become the generally accepted standard in this country among manufacturers of copper, brass, and german-silver wires, and it is this gauge that has been used in this work, unless otherwise specified. This standard has not as yet been generally accepted by manufacturers of iron wires, such as are used for telegraph lines.

191. British Standard Wire Gauge.—In Great Britain a uniform wire gauge has been adopted by law, and is now the only authorized standard in that country for all kinds of wire. The table on p. 112 covers the range of sizes ordinarily employed in telegraphy.

192. Instruments for Gauging Wire.—For quickly determining the gauge number of a wire, the ring-gauge, Fig. 72, is very convenient. It consists of a circular steel plate, having slots accurately cut in its edge, these being numbered successively from 5 to 33, covering the range of sizes of wire used in telegraph work. The smallest slot which any given wire can be made to enter shows its gauge number. Fig. 72a shows another form of gauge convenient for the pocket, in which the point at which the wire lies in the angle formed by the sides of the slot shows the corresponding number on the graduated scale by inspection. The little in-

FIG. 72. Wire Gauge—Ring Pattern.

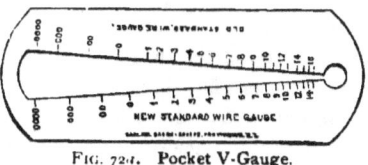

FIG. 72a. Pocket V-Gauge.

strument known as the micrometer caliper, shown of full size in
Fig. 73, is extremely accurate
and convenient. It will readily
determine the thickness not
only of wire, but of sheet-metal,
paper, or the like, from the
fraction of a mil up to 0.3 in.

193. Adaptation of Magnets to Working Currents.
—If we assume three electromagnets like that in Fig. 68,
having spools or bobbins of
equal capacity, and wind them with three different gauges of wire
(for the sake of illustration, say the three sizes shown in Fig. 69);
for each turn of a wire 1 in. in diameter we should have 4 turns of
the $\frac{1}{2}$ in. and 16 turns of the $\frac{1}{4}$ in. wire. Now, if we send a current
of 1 ampère through the thinner wire, one of 4 ampères through the
medium-sized wire, and one of 16 ampères through the thick wire,
we should find, in accordance with the principle stated in (176),
that the magnetic force would be precisely equal in each of the three
magnets. This would be true, notwithstanding the difference in
strength of current, and of thickness, length, and resistance in the
wire of the helix, because the number of *ampère-turns* is the same in
each case. We have:

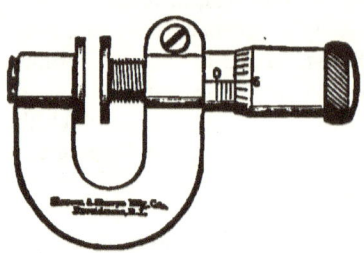

Fig. 73. Micrometer Caliper.

Diameter of Wire.	No. of Turns.	Current.	Ampère-turns.
1.00	1	16	16
50	4	4	16
25	16	1	16

A thorough understanding of this principle enables the electrician to determine the winding of his electro-magnet so as to correspond with the characteristics of the current by which it is intended
to be worked; for it will be readily seen that to produce a given
intensity of the magnetic field, upon which all magnetic effects depend, the number of turns in the coil must be in inverse proportion
to the number of ampères of current traversing the magnetizing coil.

194. Spectrum of the Electro-Magnet.—The action of the
magnetic forces in such an electro-magnet as that delineated in Fig.
68 can best be studied by means of magnetic spectra, produced in
the manner described in (68). Fig. 74 shows the spectrum of such
a magnet, when the current through the coils is barely sufficient to

Magnetic Hysteresis.

support the weight of the armature. The manner in which the magnetic circuit (177) is forced to complete itself through the air, in passing from one pole of the magnet to the other, is beautifully shown by the curving of the lines of force. When a soft-iron armature is placed in the field parallel to the polar surfaces, as shown in Fig. 75, the greater number of the lines of force are deflected so as to pass through the armature; and such being the case, the armature itself necessarily becomes a magnet of opposite polarity, whereupon mutual attraction takes place between the magnet and its armature in the direction of the lines of force which pass between them, just as if the lines were so many stretched India-rubber bands, and the force was due to their contraction. When the armature is brought into actual contact with the magnet so as to magnetically connect its poles, it becomes virtually a closed or endless core like Fig. '66, and the external lines of force in the air disappear.

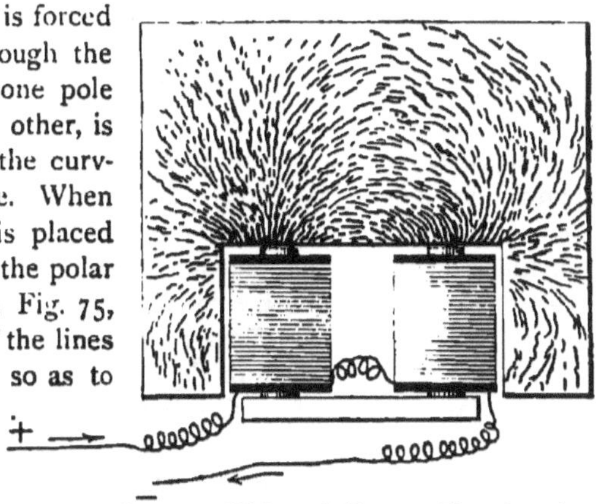

FIG. 74. Spectrum of Telegraph Magnet.—Kennely and Wilkinson.

FIG. 75. Spectrum of Telegraph Magnet and Armature.

195. Magnetic Hysteresis.—When a mass of soft iron, such as the core of an electro-magnet, becomes enveloped in a magnetic field (169), an appreciable time elapses before it acquires the maximum intensity of magnetization which the field is capable of producing. On the other hand, when the iron is withdrawn from the field, or, what is the same thing, the field is withdrawn from the iron, the latter does not lose its magnetism instantaneously; the magnetism falls off progressively in the same way in which it increased, and in almost every case some small quantity of magnetism will remain for some time, and possibly forever, after the separation of the iron from the field. This is termed *remanent*, or more commonly *residual magnetism*.[20] In some brands of cold-blast charcoal iron, when carefully annealed, such as Norwegian, Swedish, and Lowmoor iron, scarcely a trace of residual magnetism remains, and these irons are therefore preferred in the manufacture of magnet cores. Experiment has also shown that the shape of the core is no less important than its quality, and that quickness of action and freedom from residual magnetism may be best secured by making the cores as short as possible. These conditions are sufficiently fulfilled for ordinary purposes in the proportions of the magnet shown in Fig. 68, p. 91.

196. **Induction of a Current upon Itself.**—It has been stated (151) that an electric current traversing a conductor has the capacity of inducing a temporary current in a neighboring conductor. This phenomenon manifests itself in the coils of an electro-magnet in such a way that its effects are added to those of hysteresis (with which, however, they must not be confounded), so as to still further delay the magnetization and demagnetization of the iron core. These inductive effects make their appearance when the inducing current is either increased or diminished, but not while it remains steady. Further, an increasing or diminishing current not only induces a current in neighboring conductors, as indicated in Fig. 76 (in which the arrow shows the direction of the inducing current in the wire A, and of the induced current in the wire B), but it may also exercise an inductive action upon the conductor in which it flows. In a wire coiled back upon itself, as in Fig. 77, an *increasing* current, flowing in the direction of the arrow between A and B, tends to induce a current in the opposite direction between C and D, which opposes the original current and delays its increase. If, on the other hand, the current between A

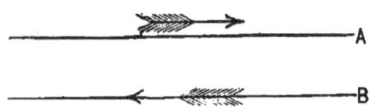

FIG. 76. Illustration of Current Induction between Parallel Wires.

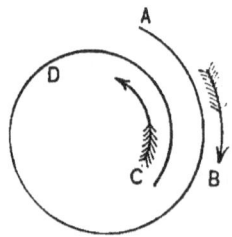

FIG. 77. Current of Self-Induction in Coiled Conductor.

---

[20] This effect was carefully studied some years since by Professor J. A. Ewing, who gave it the name of *hysteresis*, from a Greek word signifying "to lag behind," defining it as the lagging behind of changes in magnetic intensity to changes in magnetizing force. EWING: *Researches in Magnetism*, Philosophical Trans. Royal Soc., 1885.

and B is *diminishing*, it tends to induce a current between C and D in the same direction as itself, and this prolongs the duration of the original current by delaying its decrease. As the wire in the coil of an electro-magnet is placed under the same conditions as the wire in Fig. 77, it is clear that both the magnetization and the demagnetization of its core will be retarded, *first*, by the self-induction of the coil, and *second*, by the effects of hysteresis in the iron. Besides this, the presence of the iron enormously increases the normal self-induction, because the rising magnetization induces an opposing $e.\,m.\,f.$ in the wire, upon the principle explained in (78), for it will obviously make no difference whether the field be created about the wire, or whether it be moved thither from some other point in space. The sum total of these effects is termed *magnetic inertia*.

197. **Magnet Cores must not be Hardened.**—After the core of a magnet has been annealed, it is very important that it should be left black, and no attempt be made to brighten it up. If it be filed, or touched ever so little with a cutting tool, it will be slightly hardened, and will be certain to show traces of residual magnetism (195) when put to service. For the same reason the armature of an electro-magnet should never be permitted to hammer upon its poles.

198. **Effect of Self-Induction and Hysteresis in Telegraph Magnets.**—A series of experiments conducted by an officer of the U. S. Coast Survey has shown that the average period of time required for a well-proportioned telegraph magnet to release its armature, varies from 0.003 second, with maximum tension of retracting spring, to 0.033 with minimum tension.[21] The best working adjustment would be midway between these values, that is to say, 0.015 second.

199. **Other Indirect Causes of Retardation in Electro-Magnets.**—It has been stated that the magnetism developed in a given mass of iron depends solely upon two factors, the quantity of current, and the number of turns of the conducting circuit around the iron (176). It has furthermore been stated that the quantity of current traversing a circuit in turn depends solely upon the $e.\,m.\,f$ of the generator and the resistance of the conductor (127). But experiment shows that in respect to quickness of magnetization and demagnetization, irrespective of absolute intensity of magnetism, it makes a very great difference whether an exciting current of equal quantity has been produced by a low $e.\,m.\,f.$ acting through a small resistance, or by a high $e.\,m.\,f.$ acting through a proportionately

[21] G. W. DEAN: *Coast Survey Report*, 1864, p. 211.

great resistance; the magnetic actions in the latter case being far more rapid than in the former. This effect is due to the greater resistance, which in the latter case has to be overcome by the currents of self-induction set up in the coils of the magnet, which, as we have seen (196), tend, in proportion to their strength, to give rise to magnetic inertia, by delaying both the magnetization and demagnetization of the iron core. The $e.\ m.\ f.$ which tends to set up these opposing currents is necessarily of equal value in either case, as it is determined by the quantity of current in the coil and the intensity of magnetism in the core ; but the *resistance* the currents are obliged to overcome is much greater in the second case than in the first, and therefore the currents themselves are in fact very much weaker, and their retarding effect is diminished in the same proportion. This fact has an important bearing upon the working of fast-speed instruments.

200. **Electro-Magnet with Polarized Armature.**—If the armature, like the core of the magnet, is of soft iron, and placed parallel to the polar surfaces, as in Figs. 69 and 75, the action is simply one of attraction, irrespective of the polarity of the magnet, and independent of the direction of the exciting current. If, however, the armature itself be a permanent magnet (63), the direction in which it tends to move will depend upon the polarity of the electro-magnet, which in turn is determined by the direction of the exciting current.

201. In illustration of this, let the electro-magnet of Fig. 78 be provided with a *polarized armature*, consisting of a small permanent magnet *n s*, which is pivoted at one end to the yoke of the electro-magnet, while its opposite end is free to play back and forth between

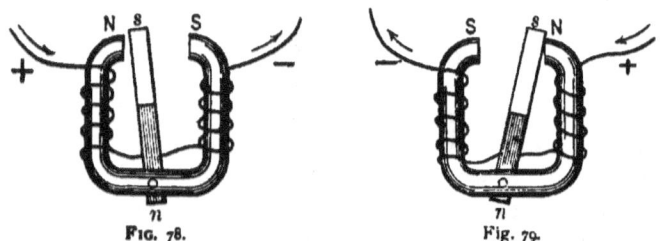

Fig. 78.   Fig. 79.

Electro-Magnet with Polarized Armature.

the poles of the N S of the electro-magnet. When the current passes in one direction, as, for example, in Fig. 78, the *n* pole of the polarized armature is attracted by the unlike pole S of the electro-magnet, and at the same time repelled by its similar pole; but upon the reversal of the direction of the exciting current, the polarity of the electro-magnet is likewise reversed, and the polarized armature is now attracted to the opposite side, as shown in Fig. 79. It is obvious, therefore, that the direction of the movement of the polarized armature depends solely upon the *direction* of the current, and not upon its *strength*. There is, therefore, an important difference between the operation of a permanently magnetic or *polarized armature* and a *non-polarized* or *neutral armature*.

202. **Combinations of Permanent and Electro-Magnets.** —Various mechanical combinations of electro and permanent magnets have been made, all of which involve essentially the same principles as the simple apparatus figured above, and by which a like effect is produced. The polarization is not necessarily confined to the armature, as similar results may be obtained by constructing the apparatus in various ways, provided that some one portion of it is polarized and another portion non-polarized. This principle is of special value in multiple telegraphy (321).

CHAPTER VII.

TELEGRAPHIC CIRCUITS.

203. It has heretofore been explained (30) that an *electric circuit* consists of an endless series or chain of conductors. That portion of the circuit which is situated between the terminals or poles, and within the generator, is called the *internal circuit*, and its resistance is the *internal resistance* of the generator; the chain of conductors which joins the poles outside of the generator is called the *external circuit*, and its resistance is the *external resistance* of the circuit.

204. The essential characteristics of every electric circuit are the same, although such a circuit may vary in length from a few inches to thousands of miles. It may be supplied with electricity from a single source, or from two or more sources situated at different points, and it may include a single receiving and transmitting instrument, or a large number of such instruments situated at different points along its course. But in every case, without regard to the length of the circuit, the time actually occupied in the transmission of the electric impulses, although not inappreciable, may be regarded, for all practical purposes of ordinary telegraphy, as instantaneous.

205. **Telegraphic Circuits.**—A telegraphic circuit is made up of the following parts: (1) the *generators*, either batteries or dynamo-electric machines; (2) the *line conductors;* (3) the *earth*, which is usually employed as a substitute for the return line wire from the distant station; and (4) the *instruments* for transmitting and receiving signals.

206. **Open and Closed Circuits.**—There are two ways in which a telegraphic circuit may be arranged for the transmission of signals. (1) The generator may be kept normally *in connection with the line*, thereby causing a constant current to traverse the circuit, and signals may be transmitted by alternately breaking and closing the circuit; or (2) the generator may be *normally disconnected from the line*, and signals may be transmitted by alternately inserting the generator into and withdrawing it from the circuit, so as to cause a current to flow for the desired period of time to form the signals.

Drawings of Electric Apparatus. 103

The first is called, in a general way, the *closed-circuit* and the second the *open-circuit* system. In other countries than North America one or the other of the above-mentioned systems is almost invariably employed, but the system in universal use in our own country, although usually spoken of as a closed-circuit system, may more properly be regarded as a compromise between the two, possessing some of the characteristics of each. As in the true closed-circuit system, the current constantly traverses the line when no work is being done, but signals are transmitted, not by interruptions of this current, but by first interrupting it at the sending point, and then transmitting the signals by closing the circuit at properly timed intervals, thus permitting the current from the generator to traverse the line and the receiving instrument, as in the open-circuit system.

207. **Drawings of Electric Apparatus.**—There are three principal methods of representing organizations of electrical apparatus: (1) by *perspective drawings*, (2) by *geometrical drawings*, and (3) by *diagrams*.

Perspective drawings are ordinary pictorial illustrations. They show the appearance of the apparatus, but, as a rule, are not well adapted to convey to the mind a clear idea of its principle and mode of operation.

Geometrical or *working drawings* consist of plans, elevations, or sections, drawn to a scale, which may represent the whole or some part of the apparatus. They usually exhibit all the constructional details, whether essential to the operation of the apparatus or not, and while indispensable to the workshop, are ordinarily of little use for purposes of explanation. Figs. 35 and 36 (pp. 46, 47) are examples of geometrical drawings.

Diagrams exhibit the apparatus, circuits, and connections, not in their actual form and proportions, but in such a conventional manner as will most clearly illustrate the principle of the apparatus and its mode of operation. Diagrams ought not to be encumbered with details which are merely constructional, and therefore unessential. The advantages of a uniform and well-understood system for the conventional representation of electrical apparatus and circuits will be apparent.

208. **Conventional Representations of Circuits and Apparatus.**—In the following paragraphs are briefly described various component parts of telegraphic circuits, with the symbolical representations which, by general consent, have been adopted to represent them, and the apparatus employed in connection with them.

(1) A wire, either straight or curved, connecting two points in a circuit. Main circuits may be in full, and local circuits in dotted lines, where such distinction is desirable.

(2) An overhead or pole line.

(3) A submarine or subterranean line or cable.

(4) The point at which any branch circuit connection is made is indicated by a round dot at the intersection. If two lines cross without being connected, the dot is omitted.

(5) In order to more readily distinguish wires which cross each other without electrical connection, it is usual to represent a loop in one of them at the crossing point.

(6) The direction of the current, from positive to negative, is shown by arrows.

(7) A waved line denotes an artificial resistance or rheostat in the circuit.

(8) An adjustable rheostat.

(9) A voltaic cell is indicated by two parallel lines, the thick line representing the zinc and the thin line the copper.

(10) The same figure arranged in the reverse way, as shown, denotes a storage battery or accumulator.

(11) A dynamo-electric machine.

(12) A ground or earth plate.

(13) A common or non-polarized relay.

(14) A polarized relay.

(15) A sounder.

(16) A recording instrument or register.

(17) A galvanoscope or galvanometer. If a tangent galvanometer, it may be represented as in Fig. 50, p. 65.

(18) A coil or loose bundle of wire, its use being indicated by a reference letter.

(19) A common Morse key.

(20) A single-current or three-point key.

(21) A single-current transmitter.

(22) A double-current transmitter.

(23) A condenser.

(24) A lightning arrester.

(25) A pole-changing switch, in which the crosses indicate the insertion of plugs.

(26) A universal switch, in which the crosses indicate points where connections are formed by inserting plugs.

(27) A three-point switch.

209. **The Earth as an Electrical Conductor.**—The earth, being composed of a vast mass of inorganic material, mostly of a porous character, and permeated throughout by water, forms an excellent conductor of electricity, and it is almost invariably employed in this capacity as a part of every telegraphic circuit. While its specific conductivity, as will appear from the table (p. 57), is much lower than that of metallic substances, yet this is abundantly compensated for by the enormous area of its cross-section.

Representations of Circuits and Apparatus. 105

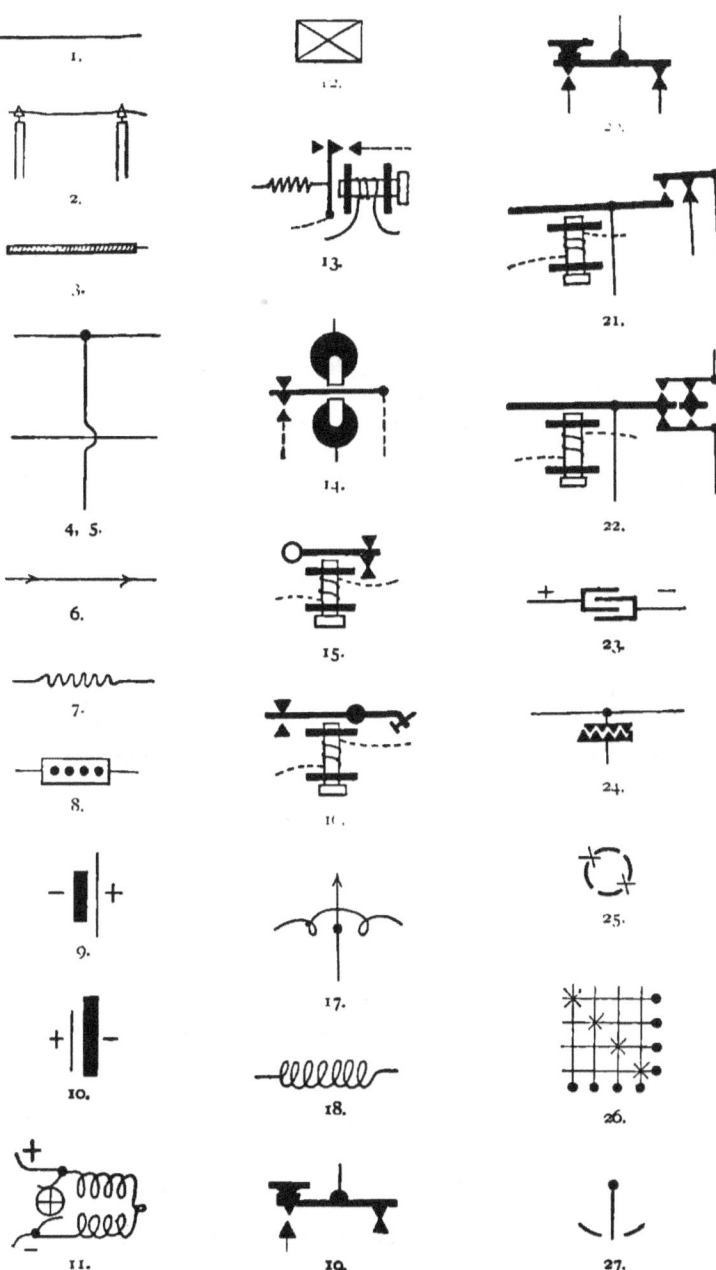

Fig. 80 illustrates the principle of the earth circuit. The current of the battery is assumed to pass through the earth from one end of the line to the other, as indicated by the arrow.

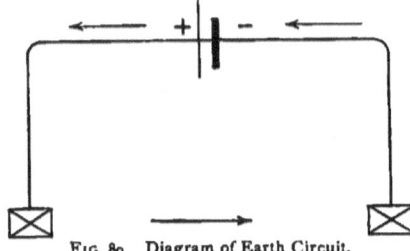

FIG. 80. Diagram of Earth Circuit.

210. **Ground Connection.**—The connection with the earth is made by means of ground-plates, which may be of sheet copper $\frac{1}{16}$ in. thick, and having an area of 36 by 48 in. Plates of galvanized iron are cheaper, and are often used instead of copper; they appear to answer the purpose perfectly well. The ground-plates should be buried in moist earth in a vertical position. In many cases an available substitute may be found by attaching the terminal of the line, by soldering or otherwise, to a pipe which forms a part of an extensive network of gas or water conductors buried in the earth, the large surface of which insures a most excellent conducting connection. It is advisable, wherever possible, to attach the wire to both gas and water pipes. When the wires are thus connected to a pipe, certain precautions are necessary to be observed, especially that of soldering the wire to the pipe outside the meter.

The connecting wire which is soldered to the ground-plate should be coated with insulating material, to prevent corrosion of the wire by the electrolytic action which might otherwise take place (27).

If circumstances render it necessary to bury a ground-plate in badly-conducting soil, as, for instance, where it is rocky, sandy, or gravelly, without sufficient moisture, a pit should be dug, and filled with scrap tin or other waste metals laid in contact with the plate, and the surface drainage and discharge from water pipes should be led into it.

211. **Advantages of the Earth Circuit.**—Several important advantages arise from the use of the earth in telegraphy as a part of the circuit. The entire cost of the return wire and its insulation is saved, while at the same time the resistance of the circuit is reduced nearly one-half. On the other hand, the inclusion of the earth materially increases the difficulty of maintaining an efficient condition of insulation throughout the circuit (219).

The specific electrical resistance of the soil and of the strata of the earth, due to the geological character of some regions, are sometimes such as to render it a matter of great difficulty to secure a sufficiently good ground connection.

An instance was observed some years since by the author in which it was impossible to secure a ground connection which would not offer an abnormally great resistance to the flow of the current. This was in the anthracite coal regions of Pennsylvania. Professor Moses G. Farmer informs him that he has met with the same difficulty in some places in the mountainous districts of New Hampshire and Vermont, on the lines between Boston and Montreal.

212. **The Open Circuit.**—A telegraph line arranged upon the open-circuit plan is illustrated in Fig. 81. Two terminal stations are shown, each having a battery, a transmitting key, and a receiving instrument. The circuit of the line divides at each key into two

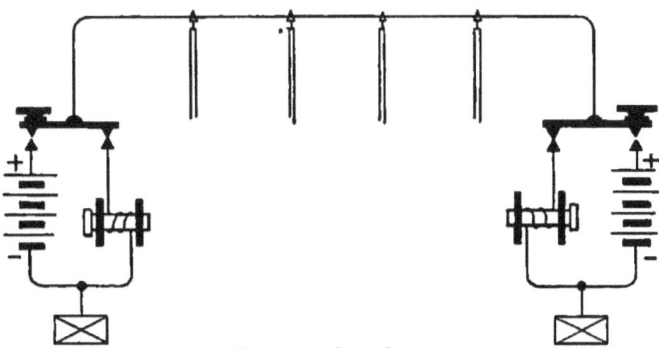

FIG. 81. Diagram of Open-Circuit System.

branches, of which only one can be closed at the same time. One branch includes the battery only, and the other the receiving instrument only. The latter branch is normally in connection with the circuit of the line. If a signal is to be sent, the key is depressed by the operator, so as to establish the connection of the line with the battery, having first broken it with the instrument. A current from the battery will now flow through the key and over the line in the direction indicated by the arrows to the other station, where it passes through the instrument contact of the key and through the receiving instrument, avoiding the battery, and thence back through the ground-plate and the intervening mass of earth to the opposite pole of the battery at the sending station, thus completing the circuit. In this arrangement, therefore, each station transmits signals by inserting its own battery at timed intervals into a circuit of conductors which is already complete.

108   *Telegraphic Circuits.*

213. **The Closed Circuit.**—Fig. 82 illustrates the closed-circuit plan, properly so called. In this the cells of the battery or batteries are always in the line, and the circuit passes normally through

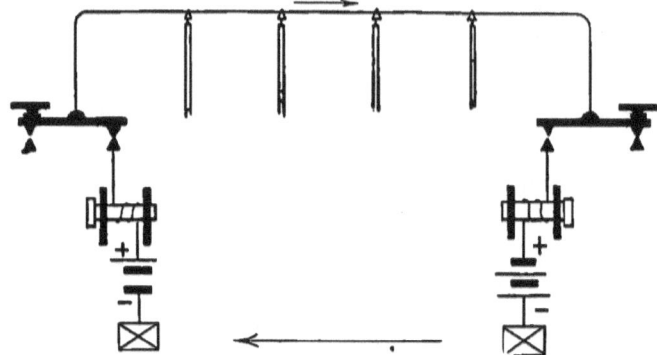

FIG. 82. Diagram of Closed-Circuit System.

the rear or breaking contact of the keys, and through the receiving instruments *at both stations*. By depressing the key at either station (as shown at the right hand in Fig. 82), the current of the entire line is interrupted, and a signal is simultaneously given upon both receiving instruments by the falling off of the armatures of the electromagnets of the receiving instruments.

214. **American Modification of the Closed Circuit.**—Fig. 83 represents the American modification of the closed circuit,

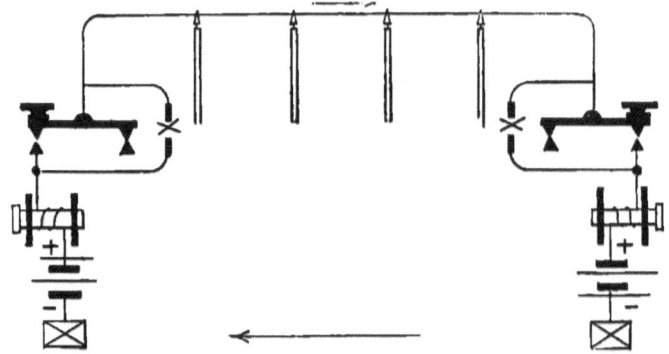

FIG. 83. Diagram of American Modification of Closed-Circuit System.

which is the standard arrangement employed in the United States, Canada, and Mexico. It differs from the last described in that the circuit does not normally pass through the key at all, but through a

Position of Battery in Closed Circuit. 109

switch or special circuit-closer beside it, which, as a matter of convenience, is in practice usually mounted upon the key, though shown separately in the diagram, as it is sometimes arranged in fact. To transmit a signal according to this plan, the circuit of the line is first broken by opening the switch, and the signals are then made by depressing the key so as to close the circuit at timed intervals upon its front contact-point. As in the last case, the alternate opening and closing of the circuit at one station affects alike the receiving instruments at all stations.

215. **Comparative Advantages of the Different Plans.**—Each of the foregoing plans of organization of a telegraphic circuit has certain peculiar advantages and disadvantages, which will be further considered hereafter. It may, however, be stated here, that one principal advantage of the closed-circuit systems is that a great number of stations may be placed upon a single line without materially interfering with each other, and may be equipped with the simplest of apparatus, all the batteries being placed at the terminal stations, where they can more conveniently receive skilled and sufficient attention.

216. **Position of Battery in Closed Circuit.**—While it is usual in a closed-circuit system to place a battery at each end of the line, as shown in Figs. 82 and 83, it is by no means an essential requirement. Comparatively short lines of say 25 or even 50 miles in length are often supplied with a battery only at one end, while very long lines are occasionally provided with an intermediate battery midway between the terminal batteries. In rare instances a battery is placed in the middle of the line only. The arrangement shown in Figs. 82 and 83 is considered preferable to any other, unless for exceptional reasons which may apply to some particular case.

217. **General Considerations respecting Telegraphic Circuits.**—In all telegraphic circuits (with the exception of those of direct working electro-chemical systems, which do not come within the scope of this work), the object sought to be obtained is to produce signals at a distant station by alternately closing and breaking the circuit at the home station, so as to alternately magnetize and demagnetize the electro-magnet of the receiving instrument at the distant station. It is therefore primarily essential that the current traversing the coils of the distant electro-magnet should be of sufficient quantity to cause the latter to attract its armature with certainty when the circuit is closed, while, on the other hand, it should be insufficient to maintain the armature in proximity to the magnet against the force of the antagonistic spring, or other retracting device,

when the circuit is broken. This result is most perfectly attained when the maximum current going through the helix of the receiving magnet is sufficient to cause the armature to be promptly attracted. and the minimum current is zero, or no current. But upon lines of ordinary length, exposed to unavoidable atmospheric influences, these conditions are usually impossible of fulfillment. The more nearly this ideal condition can be approximated to, the better are the results. It can only be fully realized upon a line of which the insulation is absolutely perfect.

218. Relation of Conductivity to Insulation Resistance.—Practically the end aimed at in all telegraphic circuits should be to make the resistance of the conductor as small as possible, and the resistance of the insulation as great as possible. Therefore, in constructing a telegraph line, it is important to employ the best possible conductor which the necessary limitations of cost will permit, and to prevent the escape of the current in undesired directions by the use of the most efficient insulators.

219. Effect of Imperfect Insulation.—The deleterious effects of imperfect insulation upon the operation of a telegraphic circuit will be understood by reference to Fig. 84, which represents

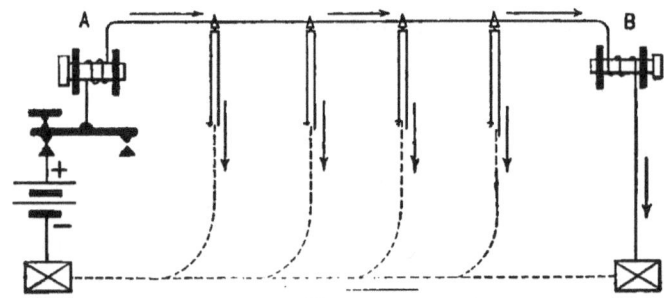

FIG. 84. Effects of Imperfect Insulation.

two stations, A and B, connected by a telegraph line, the earth being used as a return conductor. If we suppose the line to be provided with a battery at station A only, the current from its positive pole flows along the line toward B, as indicated by the arrows, but a small portion of this current escapes from the line through or across the defective insulators at every successive support. These currents of leakage find their way directly into the earth in the direction indicated by the arrows, returning to the negative pole of the battery, at A, without going through the instrument at B at all. Every imperfectly insulated point of support therefore constitutes a branch circuit (140), and causes the current to divide in proportion to the total resistance of the support and its insulator as compared with the joint resistance of that portion of the line and the branch circuits beyond the point of division. It is evident that the greater the resistance, jointly and severally, of the insulators, and the less the resistance of the line conductor, the greater will be the percentage of the total quantity of current entering the line at A which will reach the instrument at B. But as some portion of the total current must escape from the line at every point of support, it will come to pass, unless the line be perfectly insulated, that at some distance from the initial point, depending both upon the conductivity and the insulation of the line, so large a proportion of the current will have escaped from the line through the supports to the earth, that the remainder will be insufficient to produce any appreciable effect upon the receiving instrument.

220. **Working Efficiency of Telegraphic Circuit.**—The working efficiency of a telegraphic circuit is therefore determined by *the ratio between the resistance of the conductor and the resistance of the insulator.* If the total resistance of the conductor be divided by the total resistance of the whole number of insulators—that is to say, by their joint resistance—the quotient will represent the *efficiency of the circuit.* The smaller this quotient, the higher the efficiency (243).

221. **Telegraph Conductors.**—The wires used for telegraphic conductors are almost invariably either of iron or of copper. Iron wires were formerly exclusively used for outside or aerial lines. Since 1885 these have largely been superseded, in all new work, by wires of hard copper. Copper wires are invariably employed for interior work, which term comprises the wires within buildings and about the apparatus. They are also employed for all subterranean and submarine conductors. The table on page 112 gives the dimensions, weight, conductivity, resistance, etc., of the sizes of iron and copper wires most generally employed as telegraphic conductors.

222. **Iron Wires.**—Until within a few years the size of iron wire most commonly employed in the United States has been that known as No. 9, which probably still constitutes something like one-half of the total mileage of the country. Nos. 8, 6, and 4 are larger sizes which have come into use, especially since 1875. No. 4 is the largest iron wire used in this country, and No. 10 is the smallest used in the public telegraph service. These numbers refer to the so-called Birmingham gauge, and not to the American (190). See Fig. 85.

TABLE IX.
SIZE, WEIGHT, AND RESISTANCE OF TELEGRAPH WIRES.

Extra Best Best Galvanized Iron.
(Washburn & Moen Manufacturing Company.)

Gauge No. W. & M.	Diameter. Mils.	Weight, Lbs. per Mile.	Resistance. Ohms per Mile	Feet per lb.	Tensile Strength. Lbs.
4	229	730	7.	7.23	1900
6	196	540	9.5	9.59	1500
8	165	380	13.	13.89	1100
9	151	320	15.	16.50	900
10	138	268	18.	19.70	700
11	123	215	23.	24.65	550
12	108	164	32.	32.19	450
14	83	96	55.	55.	150

Galvanized Iron.
(British Post Office Specifications.)

Gauge No.	Diameter. Mils.	Weight. Lbs. per Mile.	Resistance. Ohms per Mile.	Feet per lb.	Tensile Strength. Lbs.
....	242	800	6.75	6.6	2620
....	209	600	9.	8.8	1960
....	181	450	12.	11.7	1460
....	171	400	13.5	13.2	1300
....	121	200	27.	26.4	655

Hard Drawn Copper.
(John A. Roebling's Sons Company.)

Gauge No.	Diameter. Mils.	Weight. Lbs. per Mile.	Resistance. Ohms per Mile.	Feet per lb.	Tensile Strength. Lbs.
9	114.43	209	4.3	25.2	625
10	101.89	166	5.4	31.2	525

Soft Copper.
(Geo. B. Prescott, Jr.)

American Gauge No.	Diameter. Mils.	Area. Circular Mils. (d^2)	Area. Square Inches. ($d^2 \times .7854$)	Weight and Length. Lbs. per 1000 Feet.	Weight and Length. Feet per lb.	Resistance at 75° Fahr. Ohms per 1000 Feet	Resistance at 75° Fahr. Feet per Ohm	Resistance at 75° Fahr. Ohms per lb.
10	101.9	10381	8153	31.37	31.38	1.	1000	.0313
12	80.8	6260	5128	19.73	50.69	1.59	629	.0805
14	64.1	4107	3147	12.41	80.59	2.59	386	.208
16	50.8	2583	2029	7.81	128.14	4.02	249	.515
18	40.3	1624	1276	4.91	203.76	6.39	156	1.302
20	31.9	1021	802	3.09	324.15	10.16	98	3.292

Line and Office Wires. 113

In the construction of a telegraph line, the longer each bundle or piece of wire is the better, so long as it does not exceed a weight which is convenient for the workmen to handle. Great care should be taken in making each joint, and in any case, the fewer joints the better. A loose and poorly made joint sometimes causes as much resistance as 50 miles of line.

Fig. 86 shows the common *twist-joint* most used in the United States. The ends of the wires are wrapped tightly around each other with the aid of a hand-vise and pliers, and are then *soldered*, to insure good metallic connection and to exclude moisture. The usual number of posts or supports in the United States is from 30 to 40 per mile. The smaller the number of posts the less the leakage from imperfect insulation and the less the cost.

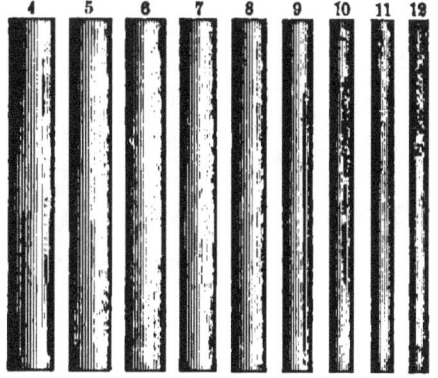

FIG. 85. Iron Wires for Telegraph Lines—Actual Size.

FIG. 86. Twist-Joint for Iron Wire.

223. Office Wires.—The copper wires used for interior wiring should generally be of No. 16 American gauge or thicker, and well covered with insulating material. If the location is perfectly dry and the number of wires is not very great, a coating of cotton braid, double, and saturated with paraffin or wax, answers very well. If there is any danger of exposure to dampness, some of the higher grades of insulated wire, most of which are known by special trade names, such as *Kerite*, *Okonite*, etc., are to be preferred. Specimens of some of the most useful varieties of these *office wires*, as they are called, are illustrated in Fig. 87. The great number of varieties of insulation now in the market offers a wide scope for selection, both in quality and cost.

Telegraphic Circuits.

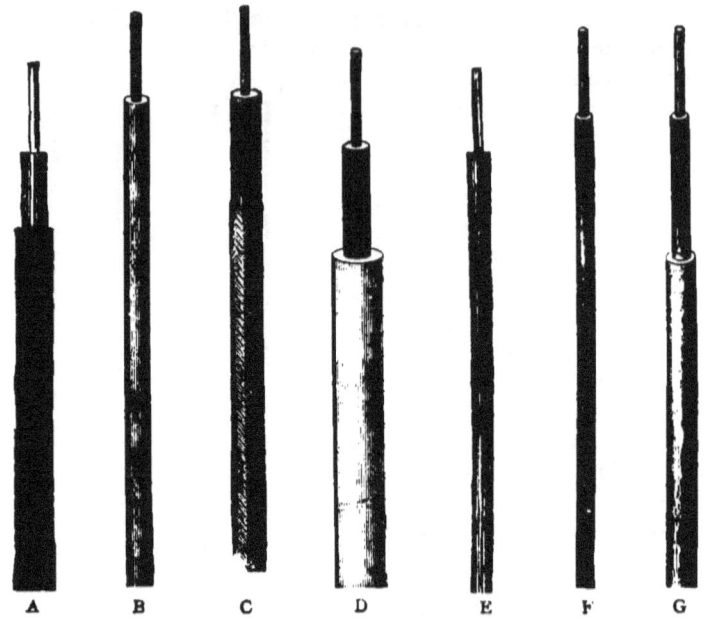

Fig. 87. Insulated Conductors for Interior Construction.

Reference Letter.	Birmingham Gauge No.	Diameter of Conductor. Mils.	Material of Insulation.	Outside Diameter of Insulation. Mils.	Outer Covering.
A	16	65	Okonite	148	Braided
B	18	49	Kerite	120	None
C	18	49	"	165	Braided
D	18	49	"	120	Lead
E	20	35	Okonite	109	None
F	20	35	Kerite	95	None
G	20	35	"	95	Lead

224. **Copper Line Wires.**—About the year 1880 it was discovered that copper wire, drawn by a process which gave it greatly increased tensile strength without materially impairing its conducting qualities, could be had in the market, and as a result many lines have since been built with this wire, with the most satisfactory results. At the prevailing prices of copper and iron, the cost of the copper line is little if any more, all things considered, than that of an iron line of equivalent conducting capacity ; while, if very great conductivity is desired, it is absolutely necessary to resort to copper, as an

iron wire thicker than No. 4 is so heavy as to be almost unmanageable.

225. Telegraphic Line Insulators.—Telegraphic lines are carried through the country supported usually upon wooden posts, but occasionally upon other structures, such as buildings, bridges, etc. These supports are separated at intervals, varying on different lines from 150 to 300 feet, or from 20 to 40 per statute mile. At each point of support each wire is affixed to an *insulator*, the office of which is to prevent, so far as possible, the escape of the current from the line through the support to the earth, in its endeavor to return to the battery by the shortest route (219). Much ingenuity has been expended, and, it must be confessed, with very unsatisfactory results, to devise an insulator which shall be capable of permanently maintaining its non-conducting properties during continued wet weather. The insulator which is in most general use in North America is an inverted cup of pressed glass, mounted upon an oak pin which forms its support, as in Fig. 87, the line being secured to its side by a *tie wire* which lies in a circumferential groove surrounding the insulator. The ordinary glass insulator is a device which has little to recommend it except its cheapness. Nevertheless, there is much to choose between the different forms in which the glass insulator is to be had. Two models in common use are shown in diametrical cross-section in Figs. 88 and 89. The figures are one-half the actual size, and the measurements are given in the drawings.

FIG. 87. Glass Insulator on Oak Bracket. Model of 1865.

226. Defects of the Glass Insulator.—The glass of which these insulators are composed is a substance which, as regards its body, is a sufficiently good non-conductor under most circumstances; but unfortunately, in rainy and damp weather, especially when the temperature of the atmosphere is rising, its entire surface becomes coated with a continuous film of moisture. This watery film forms a conductor at every support, which conveys a portion of the current from the conductor to the supporting pin upon which the insulator is mounted, from whence it finds its way into the ground, or, still worse, into some other parallel and neighboring wire. Although water is a comparatively poor conductor (116), so that the quantity of current which escapes at any one point is inconsiderable, yet, when we consider that on a line 500 miles long, there may be more than 20,000 such points of escape, the aggregate loss becomes in practice a most serious matter.

227. Resistance Influenced by Form of Insulator.—Each insulator, therefore, must be regarded in wet weather as a *conductor*, and, as such, is subject to the same law as every other conductor; that is, the resistance which it will oppose to the escaping current is directly in proportion to the length of the conducting film upon its surface, and inversely as its cross-section (118). Hence the length of the insulating surface, measured from the point of contact of the wire to the point of contact of the supporting pin, must be as great as possible. On the other hand, it is obvious that the smaller the diameter of the insulator, both external and internal, the narrower will be the conducting film, and the greater its resistance. Tested by this rule, it will be seen that the pattern illustrated in Fig. 89 must be better than that shown in Fig. 88, as in the first the actual linear distance from the conducting wire to the sup-

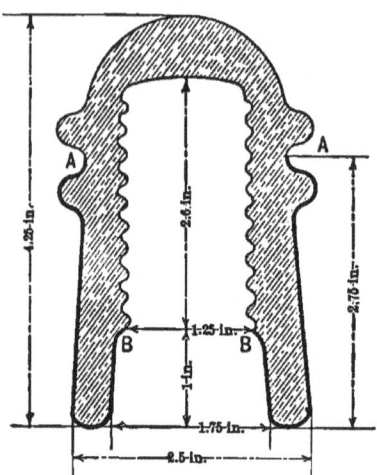

FIG. 88. Western Union Old Insulator.

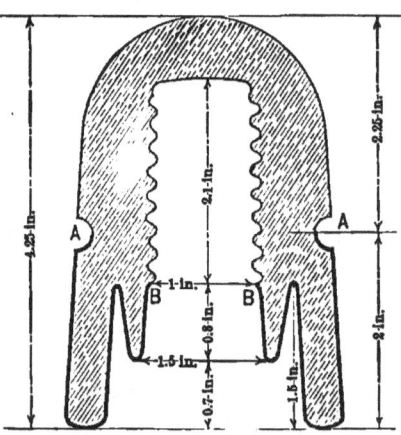

FIG. 89. Western Union Standard Insulator.

porting pin (as shown by the heavy outline) measures 5.5 in., while in the second it is only 4.3 in. It is true that the insulator, Fig. 88, is somewhat smaller in diameter than the other, which is so far an advantage; but, on the other hand, a comparatively great part of the insulating length of the latter is underneath, where it is well protected from the direct action of the falling rain.

228. **The Hard-Rubber Insulator.**—Another variety of line insulator more or less in use is the hard-rubber, which consists of a malleable iron hook for clamping and holding the wire, covered with a mass of vulcanized rubber, in cylindrical form, with a thread cut upon its exterior, which is screwed into a block, wooden arm, or other convenient support, as shown in Fig. 90.

FIG. 90. Hard Rubber Insulator. (Batchelder.)

The non-conducting properties of vulcanized rubber have been found to deteriorate very rapidly on the surface by exposure to the weather, and hence this form of insulator is now but little used except for short lines in cities, for which it possesses some advantages by reason of its small size, light weight, and general convenience.

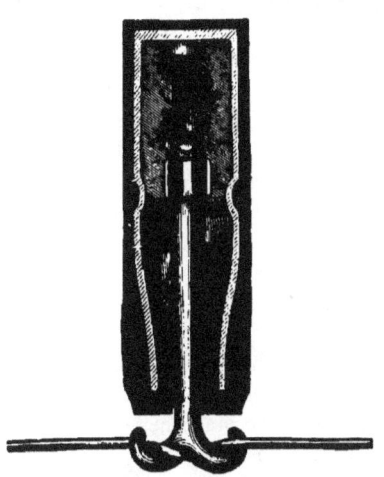

FIG. 91. Paraffin Insulator. (Brooks.)

229. **The Paraffin Insulator.**—Fig. 91 is a sectional view of the paraffin insulator, which has been much used on the railway telegraph lines of the United States. An outer cylindrical shell of cast-iron, open at its lower end, has cemented into it a narrow-necked inverted bottle of blown glass, within which again is cemented an iron stem, carrying at its lower end a hook for supporting and clamping the wire. The surface of the cement, both within and without the glass bottle, is coated with paraffin having a melting-point of about 145° Fahr. The iron shell is inserted into a hole bored in the under side of a cross-arm, which last is bolted transversely to the upright post.

230. **The Porcelain Insulator.**—The insulator shown in Fig. 92 is made in great perfection in Germany, and is extensively used in Europe, Asia, and South America, but not in the United States. All things considered, it is perhaps the most efficient insulator now known. The figure is a sectional view of the best form, known as the double bell. The material is a fine and dense porcelain, perfectly non-porous, and white in color. The glaze covers the whole internal and external surface, and is of a pure white color. The thread is smoothly formed and well-defined. The supporting bracket is of malleable iron, having an upright cylindrical stem, and the socket is packed with hemp and linseed oil when the insulator is put on. A straight iron bolt with a shoulder is used with a cross-arm, secured by a nut screwed on the under side of the arm.

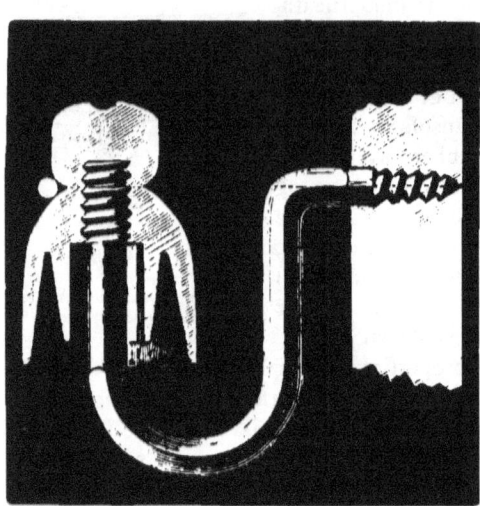

FIG. 92. German Porcelain Insulator.

231. **Defective Insulation of American Lines.**—The most serious defect in the construction of the telegraphs of the United States is unquestionably the character of the insulation. Very few of the lines exhibit any material improvement in this particular over those constructed forty years ago. It is true that the working efficiency of the more important lines has been greatly increased during the period which has since elapsed, but the improvement is due almost wholly to the use of conductors of lower resistance, and to the substitution of powerful dynamo-electric machines in the large terminal stations for the voltaic batteries formerly used. The efficiency of the less important lines is no greater, and, in many instances, not as great, as it was twenty years since. The insulators almost universally employed, as pointed out in (227), are deficient both in material and in design. In addition to their inherent defects, there are usually a considerable proportion of cracked or broken ones, which the most vigilant inspection cannot wholly prevent. The

records of the Western Union Company show that about 6 per cent. of the glass insulators on its lines require renewal yearly.[1]

232. **Effects of Climate upon Insulation.**—The combined effect of dirt and moisture upon the surface of insulators is very deleterious. Ordinary insulators in this country are affected proportionally as the air becomes charged with moisture. In the winter months this often occurs, and is notably the case when the ground is covered with melting snow, and the rain is from the south. Northeast storms begin with the wind from the northeast. Usually the wind changes to the east and south, and finally it clears up with the wind from the west and northwest. During the portion of the storm when the wind is from the southeast and south, the air is charged with moisture to its full capacity, or total saturation. It is during this time that the ordinary glass insulator is most affected. When the storm is accompanied by the wind changing in the other direction, that is, from northeast to north, and finally to northwest, the insulation is much less affected, because the atmosphere is seldom charged to over 80 per cent. of full saturation. DAVID BROOKS: *The Telegrapher*, xi. 73.

Mr. Brooks, who has devoted much attention to the investigation of questions relating to the insulation of telegraph lines, has remarked that in cities in which the fuel principally used is anthracite coal, the gas which is formed and escapes into the atmosphere produces a very deleterious effect upon the surface of glass insulators. He found while during rain, insulators in the country, in regions free from smoke, give a resistance of 60 to 100 megohms per insulator, in the city under the same conditions of weather, the resistance falls as low as 4 to 6 megohms per insulator. He instances a line in the city of Pittsburgh, a locality formerly famous for the quantity of bituminous coal-smoke which pervaded its atmosphere, where glass insulators which had been exposed on the line less than two years were so coated with soot that they gave a measured resistance of less than 1 megohm per insulator. Moses G. Farmer, who is also excellent authority in such matters, says: "I presume from long experience and many careful tests, made in the worst weather, that 9 megohms will be above the average value of three-quarters of the insulators used in this country, in the Middle and Northern States, in long-continued heavy storms."[2]

A very fair idea of the comparative efficiency of some of the different insulators referred to in this chapter may be gathered from the report of a test of five years duration, extending from March 1, 1868, to March 1, 1873. The different varieties of insulators were exposed in sets of 10, the mean resistance of this number being taken in each test. The total number of

[1] PRESCOTT: *Electricity and the Electric Telegraph*, 302.
[2] *The Telegrapher*, v. 34.

measurements, in rain, during this period was 93. The results were as follows:

Description of Insulator.	Resistance per Insulator. Megohms.		Resistance per Mile of 40 Insulators. Ohms.	
	Mean.	Minimum.	Mean.	Minimum.
1. Western Union glass, 1865 type (like Fig. 87)..........	8.3	2.6	207,500	65,000
2. Large Varley, brown stoneware with ebonite covered pin (English standard)......	8.6	3.	215,000	75,000
3. Berlin porcelain, double bell (like Fig. 92)................	28.3	19.	707,500	475,000
4. Brooks' Paraffin (like Fig. 91).	10,000.	2,300.	2,500,000,000	57,777,777
5. Boston screw-glass (nearly like Fig. 87, but with internal screw-thread), exposed 1 year	14.6	6.4	365,000	160,000
6. Western Union glass, 1871 type (nearly like Fig. 88), exposed 1 year...............	24.	3.5	600,000	87,500

The tests were made by D. Brooks of Philadelphia. *The Telegrapher:* ix. 90.

234. Distribution of Potentials in Telegraphic Circuits.—The manner in which the varying potentials at points in an electric circuit may be graphically delineated in accordance with Ohm's law, has been explained in (145). The application of this method of illustration to the specific conditions of a telegraphic circuit is instructive, as it enables the student to form, as it were, a mental picture of the electrical condition of every portion of the line when in normal condition, or when affected by leakage arising from faults and defective insulation.

In pointing out the application of this graphic method of representation, to a telegraphic circuit, it will be convenient in the first instance to assume the circuit to be perfectly insulated.

235. Potentials in Perfectly Insulated Circuit.—If a battery of 100 gravity cells in series be connected to a perfectly insulated line of say 100 miles in length, open at the distant end, as shown in Fig. 93, the line will acquire a potential throughout its entire length, of 100 volts, which is equal to the $e.\ m.\ f.$ of the battery. This will be the case, however great may be the length of the line.

236. If now the distant end of the line be connected to the earth, as in Fig. 94, a positive current will traverse the line. This will not affect the $e.\ m.\ f.$ of the battery, which remains 100 volts as

Potentials in Perfectly Insulated Circuit. 121

before, but the distribution of potentials will be changed in every part of the circuit. The distant end of the line becomes 0 or zero, being the same as the assumed potential of the earth with which it is

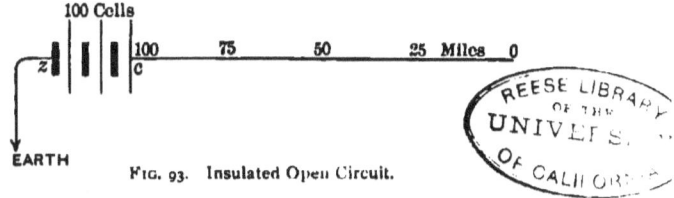

Fig. 93. Insulated Open Circuit.

directly connected, and from this point it rises gradually and uniformly along the line to the terminal or pole of the battery; at which point, as we shall hereafter see, it will be something less than 100 volts. Having ascertained the actual potential at this or any other

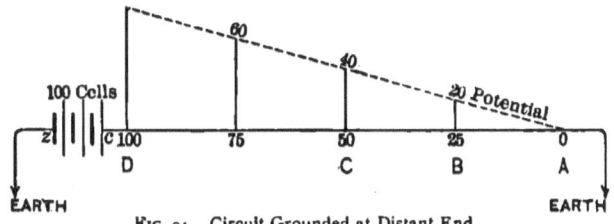

Fig. 94. Circuit Grounded at Distant End.

point on the line, it may readily be calculated for any other point, for in a circuit of uniform resistance, the potential varies directly as the distance from the zero end of the line (145). Thus if it is known to be 80 volts at 100 miles, it must be 40 volts at 50 miles,

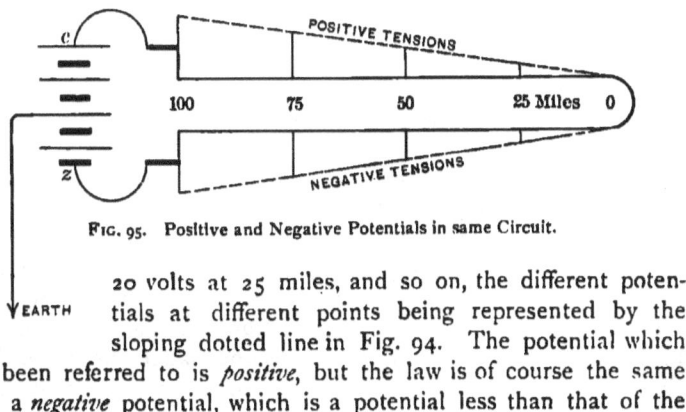

Fig. 95. Positive and Negative Potentials in same Circuit.

20 volts at 25 miles, and so on, the different potentials at different points being represented by the sloping dotted line in Fig. 94. The potential which has been referred to is *positive*, but the law is of course the same with a *negative* potential, which is a potential less than that of the

earth. An example of this is given in Fig. 95, which represents two parallel insulated lines each 100 miles in length, looped together at the distant ends. The middle of the battery c z is connected to the earth; and this point, therefore, acquires a potential of zero. The upper half of the battery imparts a positive potential to the upper line, and its lower half a negative potential to the lower line. The potential falls regularly from the c pole of the battery to o. and rises regularly from the z pole of the battery to o. If a wire were connected between the earth and any point along the length of the upper line, a current would flow from the line to the earth, the quantity of which, by Ohm's law, would be in proportion to the potential at that point. If the wire were connected in the same way to any point on the lower or negative line, a current would in like manner flow from the earth to the line. In the illustration given, it will be observed that there are *two* points of zero potential where it changes from positive to negative, one in the middle of the battery and the other at the point where the lines are looped. This is the same as the distribution of potentials in Fig. 57, p. 71, and illustrates the distribution on a telegraph line like that represented in Figs. 82 and 83, p. 108, in which there is a closed circuit with a battery at each end, these constituting electrically one battery united with the earth at its centre precisely as in Fig. 95. Fig. 96 represents a line of 100 miles connected with a battery having an

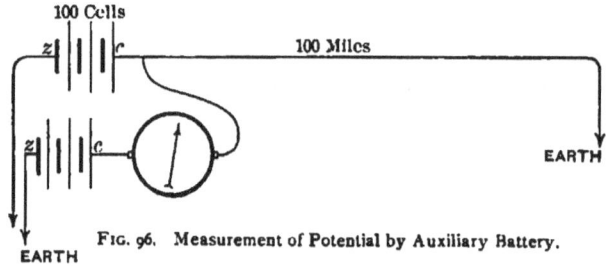

FIG. 96. Measurement of Potential by Auxiliary Battery.

e. m. f. of 100 volts, the distant end of the line being to earth. If the free pole of a second battery, with its similar pole to the earth, be now connected at any point to this line through a galvanometer, each battery will tend to send a current to the line. If the batteries are of equal potential, and attached at the same point, the needle will be deflected, say to the right. If now the number of cells and consequently the *e. m. f.* of the second battery be gradually diminished, a point will soon be reached at which no current will traverse the galvanometer, and its needle will stand at zero. When

Determination of Potential by Calculation. 123

this condition exists, the *e. m. f.* of the second battery is equivalent to the potential of the line at the point of attachment. The *e. m. f.* of the auxiliary battery will always be less than that of the principal battery, even when connected quite close to it; while as we recede from the battery, the number of cells or the *e. m. f.* required to maintain the needle at zero will gradually diminish, till, near the remote end, even a single cell will suffice to send a current into the line, because the potential at that point is approximately zero.

236. **Determination of Potential by Calculation.**—The diagram, Fig. 97, illustrates the manner in which the potential at any point on a perfectly insulated line may be calculated, when the

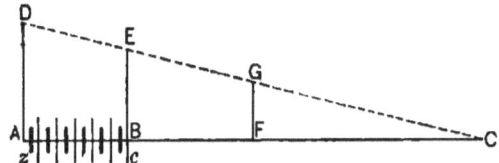

FIG. 97. Calculation of Potential from *e. m. f.* of Battery.

e. m. f. of the battery is known. Let A B represent a battery of say 100 volts, and let B C be an insulated line of any length. Let the line A B be drawn of such a length as to be in proportion to the internal resistance of the battery (131), and let the length B C correspond, in the same proportion, to the resistance of the line. Let the height of the line A D represent the *e. m. f.* of the battery. The height of the line B E will now represent the potential of the line at its junction with the battery, while the height of a vertical line, or *ordinate*, F G, will represent the potential at any other point, as, for instance, F. The potential at any point in the line may be calculated by the following rule:

As the aggregate resistance of the line and battery is to the resistance C F, measured from the distant end of the line, so is the *e. m. f.* of the battery to the potential at a given point (as F); or,

A C : F C :: A D : F G.

237. **Potentials within the Battery.**—The distribution of potentials within the battery follows precisely the same law. Fig. 98 shows a battery of 4 cells connected to a line of infinite resistance—that is, having its distant end open. The potentials are indicated by the upper dotted line, and are, under these conditions, equal to the *e. m. f.* at each point. The potential rises approximately 1 volt at each surface of contact between the zinc and the exciting solution,

the aggregate potential being attained in the fourth cell of the series.

238. Fig. 99 shows the same battery connected to a line having

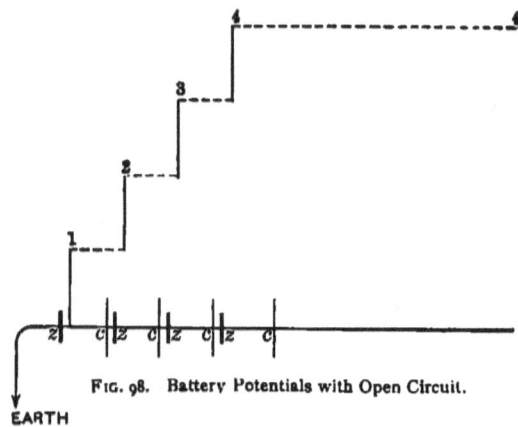

FIG. 98. Battery Potentials with Open Circuit.

the same resistance as itself; that is, the resistances of the internal and the external circuits of the battery (203) are equal. The potential at the end of the line, instead of being 4 volts, is now 0, or zero.

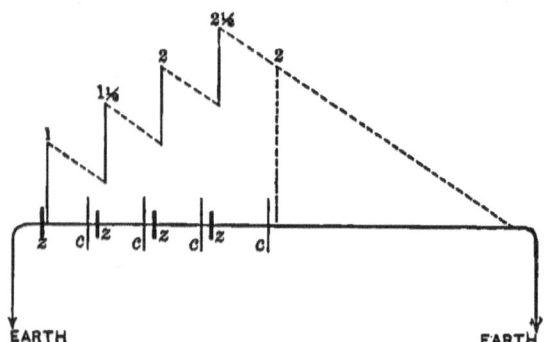

FIG. 99. Battery Potentials with Closed Circuit.

The potential at other points in the circuit, measured in volts, is shown in the diagram by corresponding figures. It will be observed that the potential falls in the liquid portion of each cell in proportion to the resistance, in the same way that it does on the line.

239. Fig. 100 shows the same battery *short-circuited*, that is, connected at each end with the earth by a wire of no appreciable resistance. The potential at both ends of the battery being now main-

Potentials in Imperfectly Insulated Circuit. 125

tained at zero, the potential rises within each cell, as in the previous examples, the maximum point of potential being at the contact of the zinc plate of each cell with the liquid, irrespective of the number of

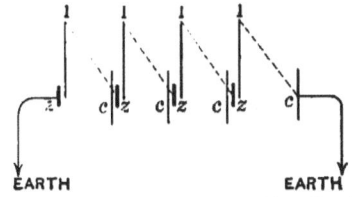

Fig. 100. Potential in Battery on Short Circuit.

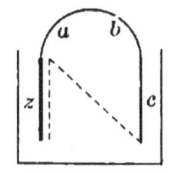

Fig. 101. Potential in Cell on Short Circuit.

cells in the series. Within each cell the potential falls as before, in proportion to the resistance of the liquid.

240. Fig. 101 shows a single cell, short-circuited by a wire ab, which is supposed to be so thick and so connected to the earth as to maintain both plates zc at a potential of zero. In this case the difference of potential in the circuit exists only within the liquid, as shown by the diagonal dotted line. In all these varied examples we find the distribution of potentials conforming strictly to Ohm's law as laid down and illustrated in Chapter V. of this work.[3]

241. **Potentials in Imperfectly Insulated Circuit.**—The distribution of potentials in a perfectly insulated circuit has now been explained, but, as a matter of fact, no telegraphic circuit is ever perfectly insulated, and in wet and unfavorable weather the

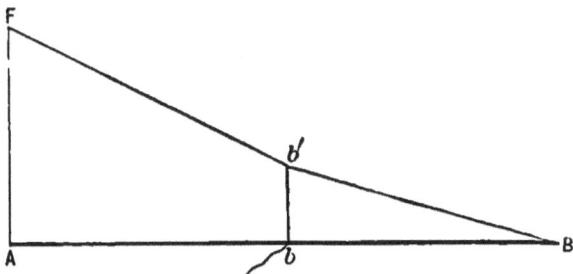

Fig. 102. Distribution of Potential on Leaky Line.

insulation is usually so defective that the normal distribution of potentials is materially modified.

[3] For the illustrations and explanations of the distribution of potentials in telegraphic circuits given in the foregoing paragraphs (234 to 240), the author desires to acknowledge his indebtedness to LATIMER CLARK'S *Electrical Measurement*, pp. 14-23; one of the very best of the early works on practical telegraphy, but unfortunately long since out of print.

In Fig. 102, let A B represent a telegraphic line, connected at A to the pole of a generator, the *e. m. f.* of which produces at that point a potential represented by the perpendicular A F. If there is situated at *b* an escape through an imperfect conductor of known resistance, the fall of potential between F and *b'*, and between *b'* and B, may be determined, provided the resistance of the line, A *b* and *b* B is known, inasmuch as it will be in proportion to the resistance (144). It will be observed that the fall of potential is greater from A to *b* than from *b* to B. Fig. 103 shows the distribution with

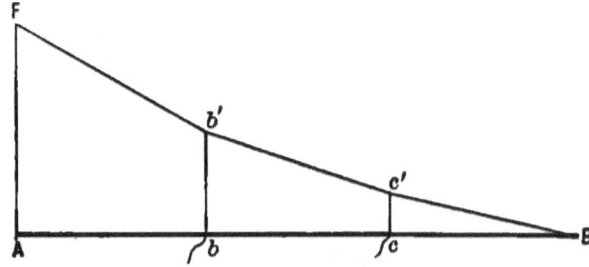

FIG. 103. Distribution of Potential on Leaky Line.

two such points of escape of equal resistance at *b* and *c*. Fig. 104, in like manner, shows the distribution with five points of escape, *b c d e f*. In each of these cases, the line at B being in direct connection with the earth, the potential at that point is zero. The *difference of potential* between each two successive points of escape becomes less and less as the distant extremity of the line is approached, and hence it follows that the quantity or effective strength of current in

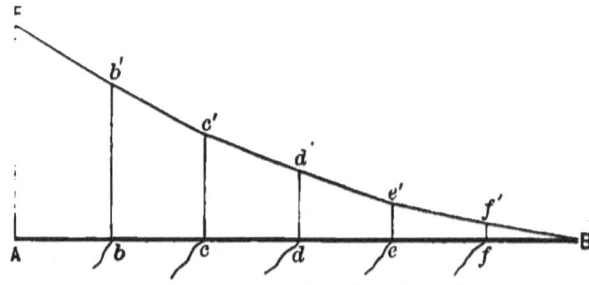

FIG. 104. Distribution of Potential on Leaky Lines

the line progressively decreases at each point, but that the decrease becomes less and less rapid as the terminal station B is approached. In an ordinary telegraph line the number of points of escape are

Effect of Imperfect Insulation upon Current.

very numerous, being necessarily equal in number to the points of support, and hence the line of potential, F B, becomes a polygon of a corresponding number of sides, or in fact a regular curve.

In Fig. 104, therefore, if the potential at the initial point of the line A is assumed at 100 volts, we might find, for example, at successive points, $b' c' d' e' f'$, the following potentials, with a perfectly insulated and with a leaky line:

Potential at point.	Insulated Line. Volts.	Leaky Line. Volts.
A	100.00	100.00
b	83.34	81.8
c	66.68	64.5
d	50.00	47.8
e	33.33	31.6
f	16.66	15.7
B	0.00	0.0

The less the resistance of the leaks, or the greater the leakage at each point, the more will the curve F B of potentials vary from the normal straight line.

242. Effect of Imperfect Insulation upon Flow of Current.—The effect of imperfect insulation upon the line, whether general or special, is to largely reduce the resistance of the line, and proportionately increase the quantity of current drawn from the batteries by the line, so that the latter are exhausted much more rapidly when the weather is wet. Hence, in working on the closed circuit plan (213, 214), the line current is *strongest* in wet weather, except near the middle of the line; but the *variation* or margin at any station, when the key is alternately opened and closed at another station, which constitutes the *working efficiency* of the line, is very much diminished. This variation or difference of course determines the available strength of signals. The effect of imperfect insulation upon the transmission of the current to a distant station has been referred to in (219). The various electrical characteristics of leaky lines have been found capable of determination by mathematical analysis.

Since, however good the insulator may be, some small portion of the current escapes from the line over it down the post to the ground, it is manifest that if the line be long, the posts many, and the insulators very poor, a small portion only of the entering current may reach the far end of the line.

The law which governs this may be thus enunciated: If the current upon the line near the battery be called the *entering current*, and that upon the distant end near where it enters the ground be called the *arriving current*, then the distance to which any stated fraction of the entering current will reach is proportioned directly to the square root of the conductivity of the wires, to the square root of the insulating power of the insulator, and inversely to the square root of the number of poles per mile used. MOSES G. FARMER: See *Report on Telegraphic Apparatus at Paris Exposition* (1867), by S. F. B. Morse, p. 69.

243. Resistance and Current in Leaky Lines.—When the average resistance of each insulator is known, it is easy to compute the *actual* insulation of the line per mile, or other unit of length. It is only necessary to divide the resistance of a single insulator by the number of insulators, inasmuch as it is simply a case of joint resistance (134). So also, when the resistance per mile of the conductor, and the resistance per mile of the insulation are both known, the *apparent* resistance of the conductor and of the insulation for any length of line may be determined. As the mathematical computations in this case are somewhat complex, a compilation of results is given in convenient form for use, in Table X, p. 129. This table shows the apparent conductivity and insulation resistance (as measured from the terminal station) of various lengths of leaky line, from 100 to 2,500 units. The table also gives the percentage of any given entering current which will reach a terminal station located at various distances along the line. Many problems arising in the working of leaky lines may be conveniently solved by means of a table of this kind.[4]

In a well insulated line, the ratio of the conductivity to the insulation resistance ought to be as low as 1 to 80,000. The table is computed upon an assumed ratio of 1 to 10,000, which is probably as much as can be relied on in rain with the most carefully constructed glass-insulated lines in our climate, and may be regarded as fairly representative of the actual present practice.

244. Computation of Working Efficiency of Line.—As an example of the use of this table, suppose it be required to determine the comparative working efficiencies of the open-circuit (Fig. 81, p. 107) and the closed-circuit system (Fig. 83, p. 108) on a line of 200 miles in length, with a conductor of 10 ohms resistance per mile, supported upon 40 poles per mile, the wet-weather value of the insulators being 4 megohms each, and the resistance of the instruments

---

[4] The author desires to express his obligations to Professor Moses G. Farmer for valued assistance in the preparation of this table. The formulas and methods of computation are discussed in the following named works: J. GAVARRET: *Télégraphie Électrique*, 376; E. E. BLAVIER: *Télégraphie Électrique*, ii. 447, 449; A. B. KEMPE: *On the Leakage of Submarine Cables; Jour. Soc. Tel. Eng.*, iv. 90; H. R. KEMPE: *Hand-book of Electrical Testing* (3d ed.), 445.

TABLE X.

RESISTANCES AND ESCAPE UPON LEAKY LINES OF VARIOUS LENGTHS.

[The unit of this table is that length of any line of which the ratio of the conductivity to the insulation resistance is as 1 : 10,000.]

True Conductivity Resistance. Line to Ground. Units.	Apparent Conductivity Resistance. Line to Ground. Units.	Apparent Insulation Resistance. Line open. Units.	Per cent. of Entering Current reaching grounded end of Line of Resistance as in first Column.
100	99.6	10030	99.5
200	197.2	5060	98.
300	291.	3430	95.7
400	379.3	2630	92.5
500	461.	2160	88.7
600	535.6	1850	84.4
700	602.5	1630	79.7
800	662.	1480	74.7
900	712.	1380	69.7
1000	760.	1300	64.8
1100	798.	1250	59.9
1200	830.	1210	55.2
1300	857.	1180	50.7
1400	880.	1150	46.5
1500	900.	1130	42.5
1600	916.	1110	38.8
1700	930.	1090	35.3
1800	942.	1070	32.2
1900	950.	1060	29.2
2000	960.	1050	26.6
2100	967.	1040	24.1
2200	974.	1030	21.9
2300	979.	1024	19.9
2400	983.	1018	16.8
2500	987.	1013	14.7
Infinite	1000.	1000	00.

100 ohms each. This would give for the line the same ratio of conductivity to insulation as that assumed in the table, viz: 1 : 10,000. The true conductivity resistance of the line from A to B (Fig. 105) is 2000 ohms. Assume the keys to be closed at both stations, the

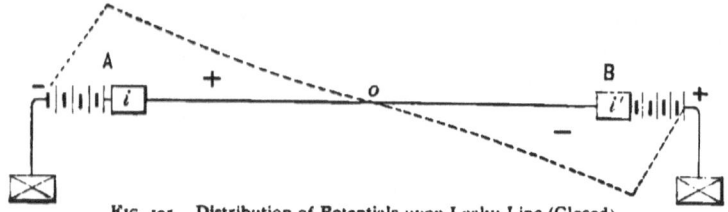

FIG. 105. Distribution of Potentials upon Leaky Line (Closed).

resistances of both instruments being alike, the e. m. f. of the two batteries also equal, and the internal resistance of the batteries to be so small, compared with that of the line, that it may be neglected. The first half of the line, from A to *o*, will be *positive*, and the other half, from *o* to B, *negative*. When sending to A, the key at B is alternately open and closed.

When key at B is open, the entire length becomes *positive*, being wholly charged from the positive pole of battery at A (Fig. 106).

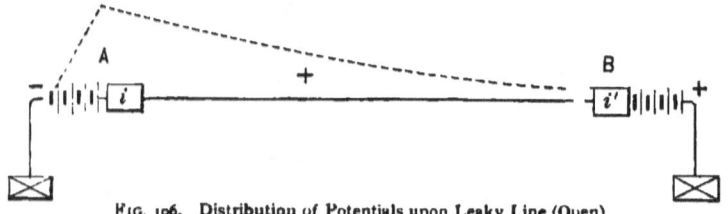

FIG. 106. Distribution of Potentials upon Leaky Line (Open).

With both keys closed, if a galvanoscope were to be connected between the point *o* and the earth, it would indicate no current, because the potential at *o* is zero. Hence the quantity of current going to line from the battery at A will be *the same as if the line were connected to the ground in the middle*—that is, at a point 100 miles from A. Assuming the e. m. f. of the battery to be 100 volts, the quantity of current entering the line at A, by Ohm's law, will be as follows:

With key closed at B:
Apparent resistance of 100 miles (1000 ohms) of
    leaky line (as per Table X.). .... 760 ohms
Resistance of instrument at A.................... 100 "
                                                                             860 "

Quantity of current 100 (volts) ÷ 860 (ohms) = .1163 (ampères).

Effect of Position of Fault. 131

With key open at B:

Apparent resistance of 200 miles (2000 ohms) of
leaky line (Table X.).......................... 1050 ohms
Resistance of instrument at A.................... 100 "
                                                  ─────
                                                  1150 "

Quantity of current 100 (volts) ÷ 1150 (ohms) = 0.087 (ampères).

Thus we have:

Current in line at A when B is closed............. .1163 ampères.
Current in line at A when B is open............. .0870 "
                                                ─────
Net efficiency of circuit..................... .0293 "

Considering next the open-circuit plan, in which the instrument at the sending station is cut out by the key while the signalling current is entering the line:

With key closed at B:

Apparent resistance of 200 miles of leaky line + instrument at A (100 ohms) = 2100 ohms................ 967 ohms
Quantity of current entering line 200 (volts) ÷ 967 ohms
= 0.206 ampères.

By fourth column in Table X, 26.6 per cent. of this only will reach the instrument at A:

26.6 per cent. of .206 is .0548. Hence:

Current in line at A when B is closed..................... .0548
Current in line at A when B is open..................... .0000
                                                        ─────
Net efficiency of circuit............................... .0548

245. This investigation shows that on a defectively insulated and leaky line, a material advantage is gained by dispensing with the battery at the receiving end of the line, and assembling it all at the sending end. In the case under consideration the difference in efficiency with the same line and instruments is as 548 to 293 in favor of the open circuit plan. With the battery constantly on the line, a computation may be made by the aid of the table, which will show that it makes no difference in the working efficiency, whether the battery be placed in equal amount at each terminal station, or whether it be all assembled in the middle of the line.[3]

246. **Effect of Position of Fault.**—The detrimental effect of a special defect in insulation or cause of escape, such as contact with the branch of a tree or a wet roof, is greatest when it is situated

---

[3] CROMWELL. F. VARLEY: *Report on Lines of Western Union Telegraph Co.* (Ms.), 1867.

midway between the terminal stations of the line, assuming the batteries and instruments to be alike at each end.

When the fault is nearer one end of the line, the station farthest from it will receive the weakest signals, and the station nearest it the strongest signals.

In increasing the battery power, in order to work over a special defect or fault in insulation, the addition should be made to the station nearest the fault.

247. **Best Position of Batteries in Circuit.**—If the insulation of a line were perfect at all times, the position of the battery in the line would be immaterial. As all lines are ordinarily subject to more or less escape or leakage throughout their length, it is obviously not advisable, except upon comparatively short lines, to place all the battery at one end; for in such case the signals will be received with much more difficulty at the station where the battery is situated, than at the opposite end of the line. The usual arrangement of a battery at each end is altogether preferable.

248. **Intermingling of Currents on Different Lines.**—The escape of the current through the insulators, poles, and cross-arms from one wire to another in wet weather, known as *cross-fire*, is a far more prolific cause of interference in the working of lines than the mere leakage to ground. This effect is sometimes miscalled induction. Weather-cross is perhaps a more appropriate term.

As electric currents always flow in greatest quantity in the direction of the least resistance, the tendency is for the currents to escape from a long circuit into a short one, or from a wire of higher into one of lower resistance. A simple escape to ground, if not too serious, may be overcome by the judicious application of increased battery-power; but when transverse leakage or cross-fire exists between different wires running upon the same line of poles, any attempt to increase the battery, in order to improve the working of one wire, produces a detrimental effect upon the working of all the others parallel to it. Upon the occurrence of a sudden shower, the effects of cross-fire are usually manifested sooner than the escape to ground, because the horizontal cross-arms are wet and become partial conductors sooner than the vertical pole.

249. **Remedy for Cross-Current.**—This difficulty may be overcome by attaching a ground wire to each pole, the upper end of which is wrapped around the central portion of each arm. These wires act to intercept the currents of leakage passing from one wire to another, and to convey them to the earth. The battery may then be increased as much as desired on any one wire without interfering

Value of Poles and Cross-Arms as Insulators. 133

with the others. It is true that the pole, even when wet, has some little value as an insulator, which is lost by this appliance, but the gain in the other direction much more than compensates for it.

250. The results of defective insulation, in causing the mixture of currents between different wires on the same line of posts, are much more detrimental near the ends of the circuits than in the middle portions; and as the terminals of the lines are almost invariably in large towns and cities where the insulation is usually much worse than elsewhere (232), the evil is aggravated accordingly. Much would be gained, therefore, by applying ground-wires to the poles even for a distance of 25 miles out from each terminus.

251. **Value of Poles and Cross-Arms as Insulators.**—Tests made many years since to determine the average resistance, in wet weather, of the pin-and-glass insulator, the cross-arm, and the pole respectively, gave some interesting results. The tests were made from New York to Philadelphia (99 miles), and to Boston (236 miles). The following, in connection with Fig. 107, will explain the method employed:

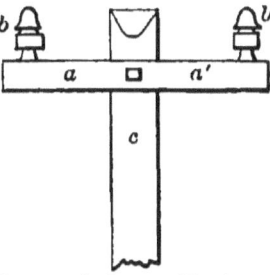

FIG. 107. Resistance of Insulators, Poles, and Arms.

Let a = the resistance of half the arm.
Let b = the average resistance of the insulator and supporting pin.
Let c = the resistance of the pole.

The following measurements were made:

(1) Resistance between No. 1 wire and the ground.
(2) Resistance between No. 1 and No. 2 wire.

The first gave $a + b + c$.
The second gave $a + a' + b + b'$.

Add together the ground tests of No. 1 and No. 2, gives

$$a + a' + b + b' = 2c.$$

By contact test $a + a' + b + b'$.
Subtract the latter from the former, and divide by 2 gives resistance of pole.

The mean of a number of tests gave for the per cent. value of $\dfrac{c}{a+b}$ of the total insulation:

Maximum.................................. .22
Minimum....... .... .11
Mean....................................... .15

The insulating power of the wet pole added to that of the insulator and cross-arm was found to be as follows:

New York to New Haven.......... 15 to 22 per cent.
New York to Philadelphia......... 11 to 12 per cent.
Mean........................... 15 per cent.

A test from New York to Boston between **two** wires, *a* and *b*, placed one above the other, as in Fig. 108, on insulators and brackets on opposite sides of the same pole, gave the following mileage insulation:

Wire *a* to ground............... 3,050 ohms per mile.
Wire *b* to ground............... 3,700   "   "   "
                                 ─────
                                 6,750   "   "   "
Wire *a* to *b* through the pole *c*.... 5,850 "   "   "
                                 ─────
                                  900   "   "   "

$$c = \frac{900}{2} = 450 \text{ ohms per mile.}$$

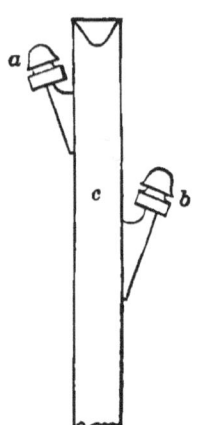

FIG. 108. Resistance of Vertical Pole.

From this it appears that the application of ground-wires to the poles would reduce the total insulation about 15 per cent., and weaken the signals perhaps 3 per cent.: but, on the other hand, it would eliminate disturbing currents amounting to about 18 per cent. of the total strength of signals.

252. Tests of Resistance of Cross-Arms.—The following measurements were made of cross-arms taken down from pole-lines in New York City. They show the insulation resistance per mile of 40 arms:

All four surfaces wet with sponge.................. 3,120 ohms.
Soaked one day, left to dry one day, and then wet... 2,680   "
Painted three years since......................... 6,150   "
Same washed....................................... 9,166   "
Very dry..................................... 11,000 to 330,000   "
Newly painted..................................... 7,214   "
Unpainted for many years.......................... 4,300   "
Same after having been well washed................13,657   "
Dry...80,000   "
Arms and pins together (wet)...................... 3,686   "

253. Tests of Glass Insulators.—Measurements made of dirty and soot-covered glass and pin insulators, taken down in New York, resulted as follows:

Dipped in water once (per mile of 40)................ 23,220 ohms.
Dipped in water 4 times (per mile of 40).............. 56,400   "
New insulator and pins direct from supply department.. 66,600   "

These figures show in a striking manner the surface deterioration of glass insulators by exposure to the smoke and dirt of a large city. Cleaning them nearly tripled their insulating power.[5]

254. Importance of High Working Efficiency.—The importance of maintaining in telegraph lines as high a ratio of insulation to conductivity resistance as possible, is shown in a striking manner by the figures given in Table X. For example, suppose it were required to determine the effect of increasing the ratio of efficiency of a given circuit from 1 : 10,000, the basis on which the table is computed, to 1 : 20,000. This might be effected, either by doubling the resistance of the insulators, or by halving the resistance of the line; that is, 8 megohm insulators might be used instead of 4, or wire of 5 ohms per mile instead of 10, either of which would affect the ratio in like manner. But the resistance of the line, referred to in (242), taken in units of the first column of Table X, is now only 1.900 instead of 2,000, and the percentage of received currents is therefore raised from 26.6 to 64.8. On a line of 250 miles, it would be raised from 14.7 to 52.9; that is, more than 3½ times as much current would be received at the end of the line.

255. Best Method of Improving Efficiency.—A line of 400 miles of No. 9 iron wire of 15 ohms conductivity resistance per mile, and carried upon glass insulators giving 4.5 megohms each in very unfavorable weather, with 30 poles per mile, would have an efficiency ratio (242) of 1 : 10,000, the same as that assumed in Table X. The true conductivity of the line would be 6,000 ohms, and the percentage of the entering current which would reach the distant end would be only 2.15. If a copper wire of 5 ohms per mile were substituted, without changing the insulation, the percentage of current received would at once be increased from 2.15 to 26.6, or more than 10 times as much, the efficiency ratio being now 1 : 30,000. The cost of the respective wires for 400 miles would be approximately as follows:

    71,120 lbs. of hard copper wire, .20 cents......... $14,344
    128,000 lbs. of galv. iron wire, .05 cents.......... 6,400

    Difference in cost.............................. $7,944

The same or a better result may be more advantageously reached by improving the insulation. If, for example, instead of the 4.5 megohm glass insulators, the German porcelain insulators of Fig. 92, p. 118, were used, the minimum resistance of which, according

[5] CROMWELL F. VARLEY: *Report on Lines of Western Union Telegraph Co.* (Ms.), 1867.

to the test, p. 120, is 19 megohms, we may safely assume that the insulation will be three times as high as with the glass. This will give us, with the 15-ohm wire, an efficiency ratio per mile of 15 : 450,000 or 1 : 30,000, as before.

The comparative cost would then be approximately as follows:

    12,000 German porcelain insulators (Fig. 92), .25 cents..... $3,000
    12,000 Best glass and bracket (Fig. 88), .05 cents.......... 600
                                                                         $2,400

It appears, therefore, that the operative value of a long line in wet weather may be increased as much by expending $2,400 in improving its insulation, as by expending $7,944 in improving its conductivity. On the other hand, it must be taken into consideration that when the line is designed to be employed for multiple transmission, a marked advantage results from high conductivity, altogether irrespective of the question of insulation efficiency. Several different strengths or values of current, in this case, require to be distinguished from each other by the selective action of the receiving instruments, and the certainty with which this can be effected depends largely upon the maximum volume of current which the line is capable of transmitting. The interference arising in fine weather from static induction (314) is also relatively much less marked upon lines of high conductivity.

256. The beneficial effects of improving the insulation on long circuits are forcibly exhibited in the following table, which contains the results of computations made by Moses G. Farmer.[6]

TABLE XI.

Distances in *miles* to which a stated percentage of entering current will reach, on a line of 18 ohms conductivity resistance per mile, with insulators of various resistances.

Per cent. of entering current received.	Insulation Resistance (Megohms per Insulator) 30 Insulators per mile.							
	1	4	9	16	36	100	1000	1600
10%	125	258	386	516	774	1290	4094	5160
25	89	178	267	356	534	890	2837	3560
50	58	116	174	232	348	580	1850	2820
75	36	73	109	146	219	365	1161	1460
90	22	45	67	90	135	235	766	900

* *The Telegrapher*, v. 269.

These results seem scarcely credible to those who have accustomed themselves to the belief that the defects of insulation on our existing lines are unavoidable, and that the only available remedy is the costly one of increased conductivity. Yet the correctness of the theory is abundantly proved by the working of such a line, for example, as the Atlantic Cable of 1866, which was 2,185 miles in length; had a resistance of about 3.7 ohms per mile, a total of over 8,000 ohms; and which worked well with an $c.\ m.\ f.$ of 10 volts, because of its high insulation. It is also a familiar fact that any good line can be worked at full speed for a distance of more than 1,000 miles in cold dry weather, when the leakage due to imperfect insulation is almost imperceptible even with sensitive measuring instruments.

CHAPTER VIII.

EQUIPMENT OF AMERICAN TELEGRAPH LINES.

257. Apparatus Essential in Telegraphy.—It has been stated that the art of electric telegraphy consists in the production, control, and organization of electric signals, which may be either visible or audible (2). The system of telegraphy now generally used in America under the name of the "Morse," produces audible signals at a distance by means of an instrument called a *sounder*, which comprises an electro-magnet; a vibrating armature; and a *key* consisting essentially of a *lever* and *contact-points*, whereby the transmitting operator is enabled to interrupt and restore the circuit with convenience and rapidity, for the purpose of forming the conventional signals.

258. Construction of the Key.—The key is made in many different forms, not essentially differing in principle. Its essential portions consist of the *lever*, the *finger-knob*, the *spring*, the switch or *circuit-closer* and the *base*. The lever, which was formerly made of cast brass, is now more usually punched from sheet steel, which renders it not only stronger but of less weight and more easy to be manipulated with rapidity.

A pattern of key much used is shown in Fig. 109. The lever A, 4 or 5 inches in length, slightly curved, is provided with *trunnions* at G, which turn between adjustable set-screws D D. The lever has a small vertical reciprocating movement upon its axis, limited in one direction by the adjustable set-screw F, and in the other by a platinum contact-point *c* inserted in a brass stud C, insulated from the frame M of the key. The finger-knob or button B, usually of non-conducting material, enables the key to be conveniently depressed at pleasure by the finger of the operator. This action brings a platinum contact-point *d*, inserted in the under side of the lever A, into contact with the one above mentioned which forms a part of the anvil.[1] One of

---

[1] Platinum is used for these and other contact-points in electrical apparatus, for the reason that the infusible properties of this metal prevent it from being oxidized by the electric spark, which tends to pass between separated conductors whenever the circuit is broken. This spark, in the case of the key, is principally due to the *inductive dis-*

Modifications of the Key. 139

the circuit-wires P is clamped to the brass rod J by means of a clamp-screw L underneath the table, this rod being in metallic connection with the base. The other circuit-wire is attached by means of another clamp-screw K to a similar brass rod I, connected with the anvil, and insulated from the frame. When the key-lever A is depressed, the circuit between the wires P and N is closed, precisely as if the wires themselves had been brought into contact with each other.

When the pressure of the finger is withdrawn, the adjustable spring E beneath the lever A, restores the latter to its normal

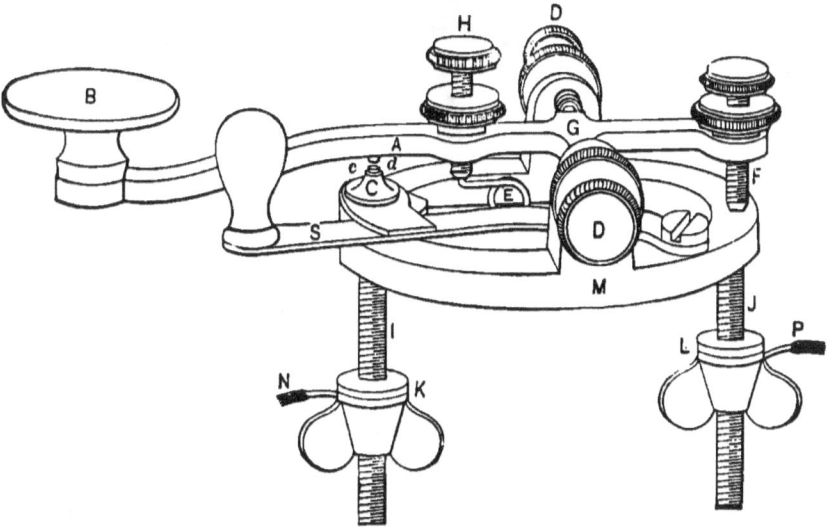

FIG. 109. Bunnell's Steel-Lever Key.

position. The upward pressure of this spring is adjusted by means of a set-screw H. When the key is not in use, the main circuit is closed by shoving the pivoted switch-lever S into a recess formed between the lip of anvil C and the frame M, thus establishing an electrical connection between the wires P and N, notwithstanding the separation of the contact-points c and d.

259. **Modifications of the Key.**—Of late years other forms of keys have been much used, in which trunnions are dispensed with. One of these is shown in Fig. 110. The lever is secured to a rear-

charge of the electro-magnets in the circuit; when there are a great number of these, the spark sometimes becomes very troublesome (196). Iridium, another infusible metal, is sometimes used instead of platinum.

wardly extending flat steel spring, the opposite end of which is firmly fastened by screws to the base. An adjustable set-screw passes through a hole in the center of the spring and its nut may be set to

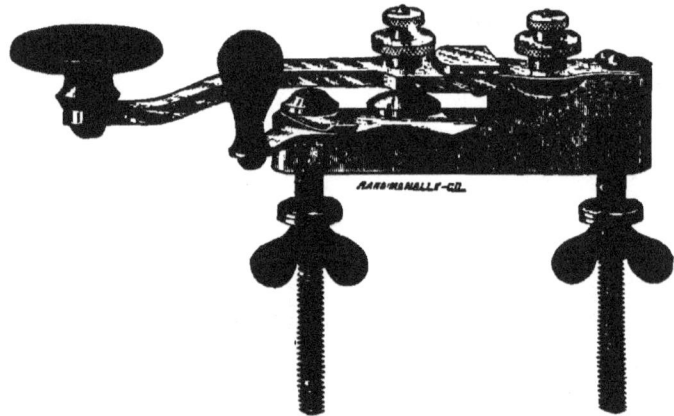

FIG. 110. Western Electric Key.

bear upon its upper surface, thus enabling its flexibility to be regulated as desired. A second check-nut is capable of adjustment to regulate the stroke, or extent of vertical vibration of the key-lever.

A modification which is applicable to all keys, consists in placing two binding-screws, one of which is insulated, upon the top of the base as in Fig. 111, to which the wire connections are made, thus avoiding the necessity of boring holes through the table, which is sometimes objectionable.

FIG. 111. Victor Key.

260. **Adjustment of the Key.**—In adjusting a key for work, the best result will usually be attained by giving the lever a small movement with a moderate upward spring-pressure. Trunnion keys should be carefully adjusted, so as to prevent unnecessary lateral movement on the one hand, and unnecessary friction on the other.

The Sounder. 141

A trunnionless key, working upon knife-edged bearings, now very extensively used and known as the "Victor," is shown in Fig. 111. By this device, friction and weight are reduced to a minimum, while adjustment is rendered more convenient. In this and in the preceding pattern of keys, the electrical contact being made through continuous metal, is more perfect than is possible when trunnions are used.

261. The Sounder.—The essential parts of this instrument, shown in outline in Fig. 112, are an *electro-magnet* E, usually about

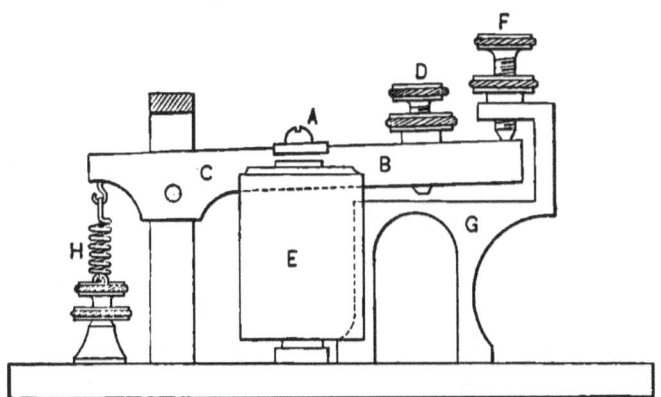

FIG. 112. Elevation of Sounder.

the size and proportion of that shown in Fig. 69, page 91, and an armature A fixed transversely upon a brass lever B about 3 inches in length, fitted with trunnions at C and mounted between transverse set-screws in the same manner as the key. Two other set-screws D and F form adjustable stops, which limit the vertical motion of the lever in each direction. When the circuit is closed through the electro-magnet, the armature is strongly attracted, and is thereby made to strike forcibly against a sounding-post or bridge G through which the vibration is imparted to the table upon which the instrument is secured. When the magnetism disappears, the lever is thrown against the upper stop F by the recoil of an adjustable spring H. The operator interprets the signals by mentally noting the difference in character between the sounds of the *down-stroke* and the *up-stroke*, and by estimating the space of time intervening between them, as will be hereinafter explained (378). Fig. 113 is a common pattern of sounder, about two-thirds the actual size. The helices of the electro-magnet are wound with insulated wire, the **thickness and convolutions of which depend upon the strength of** current with which the instrument is designed to be used. When intended for *direct working*, in which case the sounder is actuated by the current received over the line from the distant station, the helices are wound with wire which may differ in gauge from number 22 to 32 and even 36 (see table, page 94) according to the length or resistance of the circuit in which they are intended to be used. Ordinarily the sounder is operated by a *local battery*, consisting of a single gravity cell (Fig. 4, page 5), in which case its electro-magnet is wound with number 24 wire to a resistance of about 4 ohms.

FIG. 113. Sounder.

262. **Short Line Instrument.**—When the length of the line on which the sounder is to be used does not exceed 40 or 50 miles, a convenient and compact form of apparatus, consisting of a sounder and key mounted on one base, with proper electrical connections as shown in Fig. 114, and having its electro-magnet wound to a resistance of 20 or 30 ohms, may be employed with advantage.

263. **Adjustment of the Sounder.**—The adjustment of the sounder may be best effected as follows:—*First*, loosen the stop D until the armature A rests directly upon the poles of the electro-magnet. *Second*, set the trunnion-screws upon which the lever B turns, as tightly as possible without in the least binding the axis. This can be determined by lifting and letting fall the lever, having

Pocket Apparatus.

previously slackened the spring H. It should drop freely when released. *Third*, lay a piece of paper between the poles of the magnet and the armature, and close the circuit through the magnet,

FIG. 114. Combination Sounder and Key.

so that the attraction exerted upon the armature will clamp the paper. *Fourth*, adjust the screw-stop D, until the armature is raised just enough to permit the paper to be drawn out without friction. *Fifth*, adjust the stroke by means of the screw F.

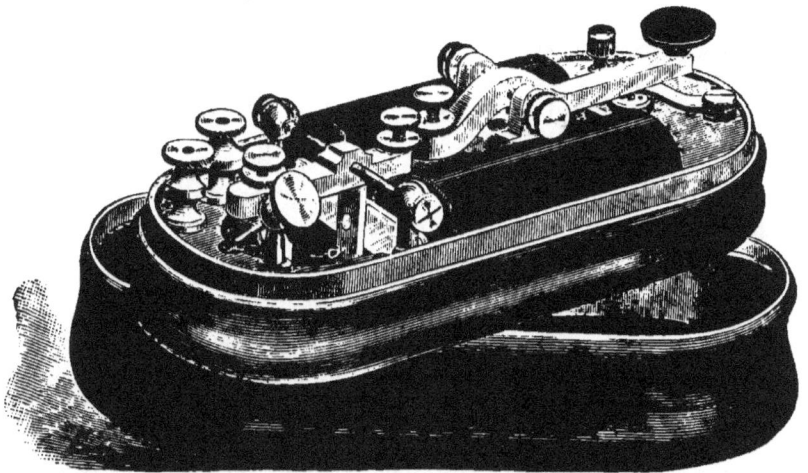

FIG. 115. Pocket Sounder and Key.

Sixth, strain the retracting spring H until the character of the sounds produced by the up and down strokes of the lever is satisfactory.

264. Pocket Apparatus.—Fig. 115 represents a convenient, compact, and exceedingly portable form of direct-working sounder, having a key attached, so as to form a complete apparatus for sending and receiving communications. The engraving is nearly the actual size of the instrument, which, together with its case, weighs but a few ounces, and can be readily carried in the coat-pocket.

265. Box Sounder.—Fig. 116 shows still another combined key and sounder, in which the electro-magnet is of standard dimensions, and is enclosed within a wooden box, the resonance of which materially increases the volume of sound made by the strokes of armature-lever. This apparatus, being complete in itself, is often found useful in the railway telegraph service, for establishing

FIG. 116. Combination Box Sounder and Key.

temporary communication at any point along the line in case of accident.

266. Working by Relay and Local Circuit.—When the line is of considerable length and corresponding resistance (118), or its insulation is defective, as is usually the case in practice (232), the main line current may be too feeble or too variable to satisfactorily operate the receiving instrument. This inconvenience is avoided by making use of an intermediate receiving instrument called the *relay*, which is included in the main circuit. Its armature-lever has only to perform the function of opening and closing the circuit of a local battery at the receiving station, by which means the sounder can be made to produce any required volume of sound.

267. Construction of the Relay.—The relay consists of an electro-magnet, having its armature delicately poised, so as to be free and capable of being acted upon by minimum magnetic attraction. Fig. 117 represents a design which is largely used.

Fig. 118 represents the working parts of a relay in outline. The electro-magnet M has its soft iron cores, usually 2 in. long and

Construction of the Relay. 145

1½ in. diameter, screwed into a yoke Y, 2 in. long and ¼ in. in thickness. The standard resistance of the coils is about 150 ohms, and the average number of convolutions of wire in the coils 8,500. The

FIG. 117. Western Union Relay.

usual magnetizing force is about 200 ampère-turns (176). The front ends of the magnet are supported in a vertical metallic frame F, the foot of which is firmly secured to a hard-wood base, by screws

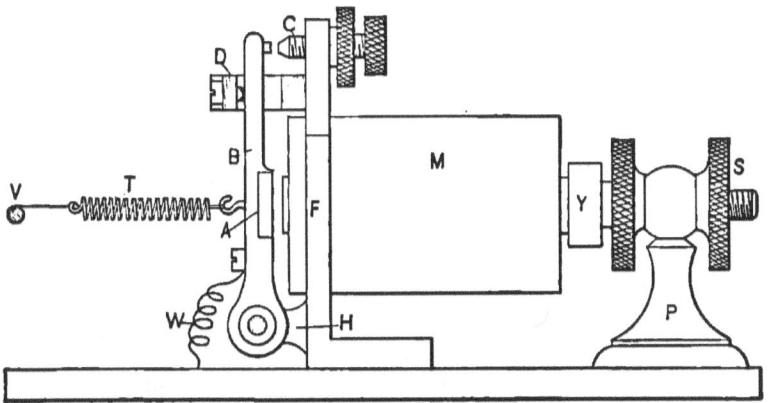

FIG. 118. Elevation of Relay.

from beneath. Two circular openings are formed in this frame, large enough to permit the helices to pass freely through without being fastened in any way. The yoke end of the magnet is supported at

S by a right and left screw, or a straight screw with two check-nuts passing through a brass pillar fixed upon the base. This device is capable of imparting to the electro-magnet M a limited horizontal advance or retrograde movement. The armature A in front of the poles is fixed transversely to the upright lever B, the lower end of which is mounted upon a steel arbor turning between two adjustable set-screws, mounted upon standards H projecting from the lower part of the frame F. The armature-lever and armature are permitted a limited movement to-and-fro upon the axis, responsive in one direction to the attraction of the electro-magnet, and in the other the retractile force of the spiral spring T. This motion is limited in one direction by the adjustable screw-stop C, and in the other by a fixed stop of non-conducting material placed within the slotted projection D, through which the lever B passes freely, not touching anything but the stops. The spring T is attached at one end to the lever B by a hook, and at the other end to a thread which winds upon a spindle V provided with a milled head. (See Fig. 117.) This spindle is supported in a socket upon the end of a horizontal brass rod, which slides through a pillar and may be fastened in any required position by a check-screw. The object of this device is to enable the tension of the spring T to be adjusted through a somewhat wide range, the necessity for which will be hereinafter explained.

The electrical connections of the relay are as follows :—Upon the base are four binding-screws for the attachment of wires. Only three of these are visible in Fig. 117, the other being concealed by the electro-magnet M. The insulated copper wires projecting from the helices of the electro-magnet are seen to pass down through the wooden base, underneath which they are connected to the two right-hand binding-posts, so that current entering at one binding-post, after traversing both helices of the electro-magnet, returns to the other and passes on.

It has been explained that the function of the relay is simply to break and close the independent *local circuit* in which the sounder is included, whenever the main circuit is broken and closed, or in other words, to repeat the signals of the main circuit into the local circuit. To this end the armature-lever A is carefully insulated from the frame F by a non-conducting bushing interposed between the lever and the axis upon which it turns. A wire leads underneath the base from one binding-screw to the support of the armature. Another thin copper wire W, coiled into a spiral, connects the armature-lever with its support, being attached to the former by a small screw seen in the figure just above the axis. When the armature is attracted by the electro-magnet, a platinum contact-point near the top of the lever B (Fig. 118) is brought in contact with a corresponding platinum point which forms the tip of the adjustable screw-stop C. The stop C is in electrical connection with the brass frame F, and this is in turn connected with the extreme left-hand terminal binding-post by a wire under the base. Hence it necessarily follows that whenever the two platinum points are brought in contact by the advance movement of the armature-lever in response to the attraction of the electro-magnet, a connection will be formed between the two terminals completing the local circuit through the sounder.

268. Adjustments of the Relay.—The adjustments of the relay are three in number: *first*, the stroke, or extent of the to-and-fro movement of the armature; *second*, the antagonistic action of the retracting spring; and *third*, the distance between the armature and the poles of the magnet.

The first-mentioned adjustment is effected by the front screw-stop C, which is movable, the rear stop being fixed. The maximum separation between the platinum contacts ought never to exceed $\frac{1}{32}$ of an inch, and in case the actuating current is weak, it should be made as much less than this as possible. Under ordinary conditions this adjustment, once properly made, seldom requires alteration. The second adjustment particularly requires judgment and skill on the part of the operator. When the attraction of the electro-magnet is very strong, as is usually the case in wet weather (242), the armature will not fall off promptly, and hence the tension of the spring must be increased by turning the milled head, thus winding the thread and straining the spring. The spring T must never be wound around the spindle V. When the thread has all been taken up, the spindle must be removed to a greater distance from the armature, by loosening the check-screw and sliding the rod upon which it is mounted through the post to a sufficient distance and then clamping it again. In extreme cases the tension cannot be sufficiently increased by this means, and it then becomes necessary to resort to the third adjustment, which consists in withdrawing the magnet by the screws, so as to increase the distance between the poles of the electro-magnet and the armature.

269. The Register.—This apparatus for recording the signals, was originally regarded as an essential part of the Morse system. It is now but little used except at small stations and on railway lines. It consists essentially of a pair of grooved rollers moved at a

148 *Equipment of American Telegraph Lines*

uniform rate by a system of controlled clock-work driven by a weight or spring. A long narrow strip or ribbon of paper, taken from a roll, passes between adjustable guides, and thence between the grooved rollers, the motion of which draws it along from right to left at a uniform rate. An electro-magnet is provided with a lever similar to that of the sounder (261), and armed at the extremity with a style or point of hard steel, which works in a groove in the upper roller. As the strip of paper passes between the rollers, a raised line is embossed upon the upper surface of the paper whenever the marking point is forced into the groove by the attraction of electro-magnet exerted upon the armature at its opposite end of the lever. A retracting spring is provided with adjustments similar to those described in connection with the sounder. The paper-guide is capable of lateral adjustment, so that the same strip of paper may, if desired, be used several times, each successive line of characters lying parallel to, and in front of the preceding one.

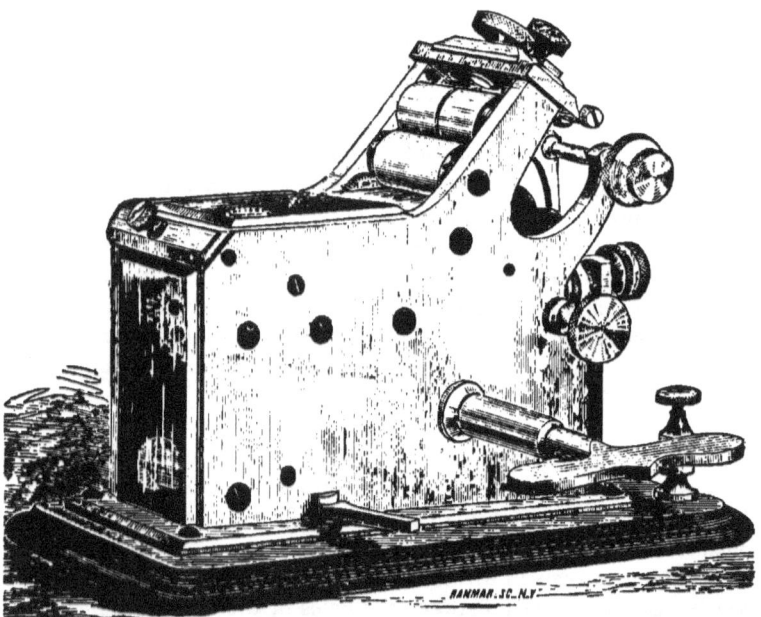

FIG. 119. European Pattern Register.

270. Fig. 119 shows one of the most modern forms of the register, in which the clock-work is propelled by a coiled spring, instead of by a weight, as in the instruments formerly made. The ma-

chinery is entirely enclosed in a brass case with plate-glass panels, so as to exclude dust. The end of the strip of paper is inserted through the guide and between the rollers while the clock-work of the register is in motion. The rate of speed at which the paper moves may be varied, within certain limits, by means of a governor acting

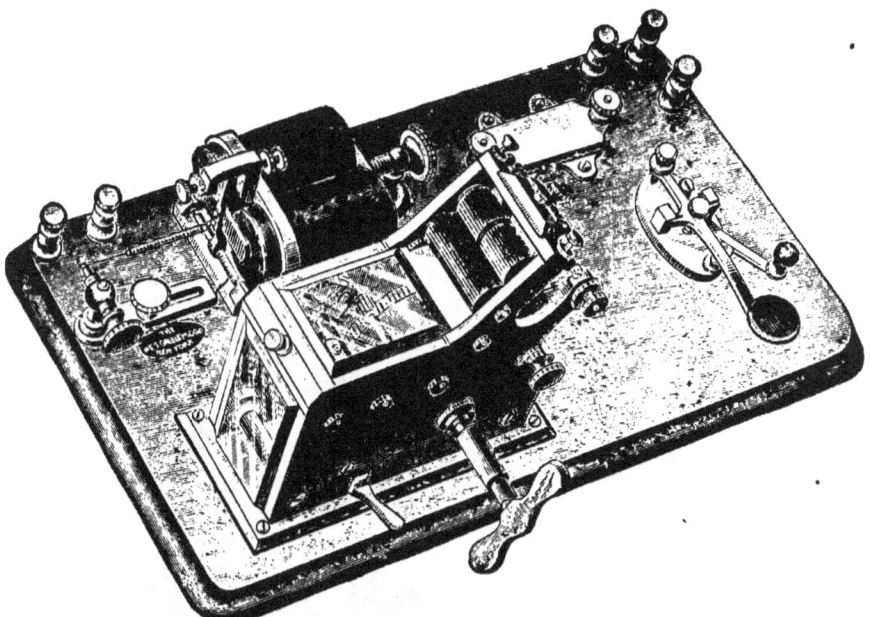

FIG. 120. Combination Victor Key, Relay, and European Register.

upon the retarding or controllling device of the clock-work. The register, relay, and key, are sometimes combined upon a single base with their necessary connections, as in Fig. 120.

271. **Adjustments of the Register.**—The armature must be so adjusted as not to come quite in contact with the poles of the electro-magnet (262). After the set-screw which limits the movements of the armature toward the electro-magnet has been properly adjusted, the armature should be held down, either by closing the local circuit through the electro-magnet or by the finger; the register is then started, allowing the paper to run, and the marking-point adjusted, by turning its milled head until a continuous uniform line is produced upon the upper surface of the paper, by the action of the style between the groove and the roller. The embossed line should not be made any deeper than to enable it to be distinctly seen when

placed in a transverse light in front of a window. The final adjustment is that of the set-screw which limits the movement of the armature away from the electro-magnet, which should be so fixed that the marking-point will just clear the paper when released by the breaking of the circuit.

272. **Causes of Defective Marking.**—Imperfect marks will result if the style does not work accurately in the groove of the upper roller. This defect is usually due to the working loose of the screws which form the transverse bearings of the axis of the marking lever, so as to permit too much lateral play, and may obviously be remedied by proper adjustment (260). When such adjustment has been effected it should be let alone. The unnecessary alteration of the adjustment of the pen-lever is frequently the cause of much trouble to inexperienced operators.

273. **Ink-Writing Register.**—A form of register now much used is shown in Fig. 121, in which the characters are marked upon the upper surface of a narrow strip of paper by means of a small sharp-edged jockey-wheel, driven by the clock-work and supplied with ink by means of a felt roller saturated with thick, oily ink, which revolves in contact with it, being

FIG. 121. Ink-writing Register.

held thereto by a spring. A knife-edge on the end of the armature-lever, lifts the slack of the paper into forcible contact with the under edge of the revolving inked jockey-wheel, whenever the armature is attracted, and thus the characters are plainly recorded in black ink.

274. Circuits of the American System.—Telegraphic circuits in the United States and Canada are almost universally arranged upon the modification of the closed-circuit system, shown in Fig. 83, page 108. If an operator at any station desires to transmit a communication, he first opens the switch of his key and interrupts the flow of current throughout the line. This causes the armatures of all the relays in the circuit to fall off. He then proceeds to manipulate his key, by closing and breaking the circuit at accurately timed intervals, so as to form the conventional characters of the telegraphic code in the order required to spell out his communication. During this operation, the armatures both of his own and of all the other relays in the circuit, as well as those of the registers or sounders connected therewith, will respond instantaneously to every movement of the key, and consequently the communication may be copied from the sounder or register by an operator either at any one station, or if required, at all the stations simultaneously. As the receiving instrument of the sending operator normally responds to every movement of his key, it is evident that the receiving operator at any station may interrupt him at any time, by opening his own key, and thus breaking the circuit in another place. The sending operator instantly perceives such an interruption, by reason of the failure of his relay and sounder to respond to the movements of his own key.

275. Arrangement of Apparatus at a Way-Station.—The simplest complete combination of apparatus is that found at an intermediate or way-station having but a single main wire. It comprises one set of instruments, viz: a key, sounder or register, relay, local battery, switch, lightning arrester, and the wires connecting the different parts of the apparatus. The manner in which these are usually arranged and connected with each other, and with the line, will be understood by reference to Fig. 122, which represents a complete way-station.

The sounder, relay and key may be conveniently placed upon a table about two feet by three in size. The sounder, or its equivalent the register, is best placed in the middle of the length of the table, having the key on the right and the relay on the left. The knob of the key should be about 12 in. from the front edge of the table. The switchboard is most usually placed in an upright position upon the wall above the table, or in any convenient position. The switchboard is not absolutely necessary, but it is a very convenient device

152 Equipment of American Telegraph Lines.

for making the necessary changes in the connections of the wires. In the form most commonly employed (see Fig. 131) a device is used similar to that described in connection with the rheostat (Fig. 43, p.

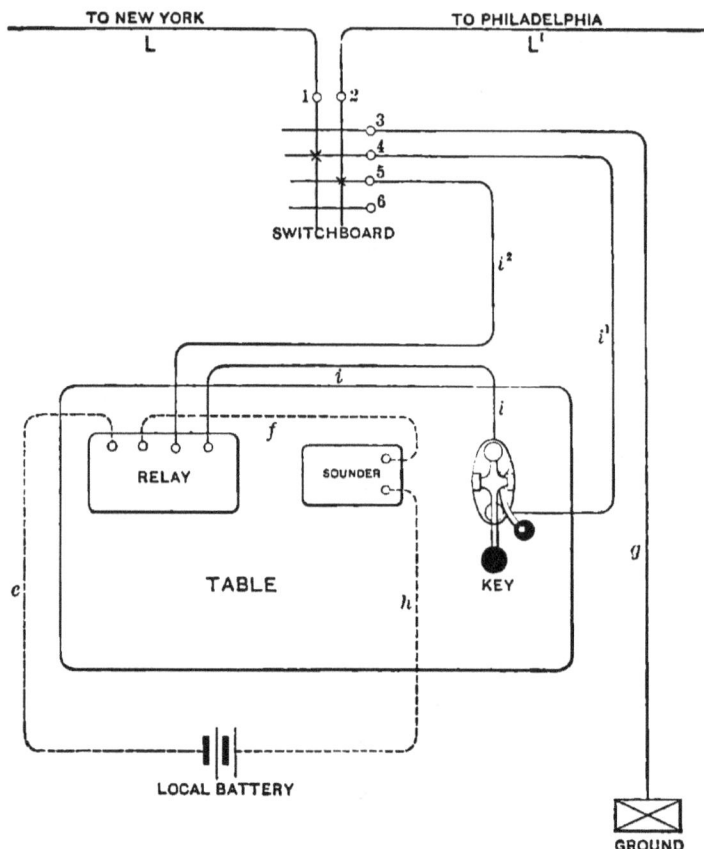

FIG. 122. Diagram of Way-Station.

51), a number of metallic pegs with insulating handles adapted to be inserted between the edges of thick plates of brass, between which they form an electrical connection.

276. Connections of Apparatus of Way-Station.—The way-station in the diagram Fig. 122, may represent for instance, Trenton, N. J.; L being the line-wire from New York to Trenton, and L^1, the line-wire from Trenton to Philadelphia. The line-wires are extended into the station building by *leading-in wires* properly insulated, which are connected with the binding screws 1 and 2 of the switch. These binding screws are mounted upon vertical metallic bars secured to the wooden back-board see Fig. 131). Of the instrument wires i^1 leads from the binding screw 4 of the switch to one of the terminal screws of the key. From the other terminal screw of the key, the wire i extends to the first right-hand main terminal of the relay; the other instrument wire i^2 unites the binding screw 5 of the switch to the second right-hand terminal of the relay ; a wire g leads from the binding screw 3 of the switch to the ground (210). The wires of the local or office circuit are run as follows :—From the local battery, the wire e runs to the first left-hand (local) terminal of the relay ; the wire f runs from the second left-hand (local) terminal of the relay to one terminal of the sounder or register ; and finally the wire h runs from the remaining terminal of the sounder or register to the other pole of the local battery. It is quite immaterial which pole of the battery is connected to the relay and which to the sounder.

277. **Manipulation of the Switchboard.**—The various changes which may be made upon the one-line switch for different purposes are as follows :

(1.) *To cut out the Apparatus.*—Insert pegs connecting 1 and 2 with 6, as in Fig. 123. The main-line current now passes from the + pole of the main battery at New York through the instrument at that station, and thence over the line L to 1, and through 6 to 2, thence over the line L¹, and through the apparatus at Philadelphia to the — pole of the main battery at that place, and thence to the ground. The — pole of the New York battery is also connected with the ground, and therefore the circuit is complete. When thus

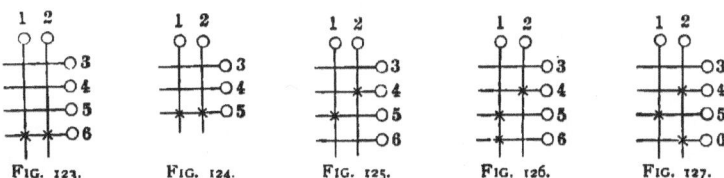

FIG. 123.   FIG. 124.   FIG. 125.   FIG. 126.   FIG. 127.

arranged, it will be seen that the ground wire g, and also the instrument wires i^1 and i^2 at the way-station, are entirely disconnected from the main line.

This is the manner in which the pegs of the switch should always be arranged when the operator is absent from the office or during the prevalence of a thunder-storm.

In some of the switches in use, the cut-out bar 6 is omitted When this is the case, the holes on bar 5 should both be pegged, and the others left open (Fig. 124). The key (Fig. 122) should also be opened, which serves to disconnect the wire i^1; i^2 being open at the switch. This prevents lightning from injuring the instruments. In thus *cutting-out* by means of one of the instrument wires, *the one which runs to the key* should always be made use of, not the one going to the relay.

(2.) *To cut in the Apparatus.*—Insert pegs in 1-5 and 2-4 (Fig. 125) or else in 1-4 and 2-5 (it is immaterial which, though the former is most usual), and remove the pegs from 1-6 and 2-6, taking care that the remainder of the holes are also open. It is better to peg holes 1-5 and 2-4 before taking out 1-6 and 2-6, which can readily be done if four pegs are provided, as then the circuit of the main line need not be interrupted even for an instant. Care should always be taken before "cutting in" the instruments to see that the key-switch is closed.

If Nos. 1-5 and 2-5 are used for cutting out, close the key, insert peg at 1-4, and afterwards remove peg from 1-5.

278. **Testing for Disconnection.**—In case the line is broken, or the *circuit is open*, as it is termed, it becomes necessary for the way-station to test the circuit. This is done by *grounding the line*. Suppose the line disconnected at some unknown point. Trenton already has pegs in the holes 1-5, and 2-4 (Fig. 125), as in the ordinary manner of working, but of course perceives no current. A spare peg is placed in 1-6 (Fig. 126), the effect of which is to connect the end of the New York line L with the earth wire g. This completes the circuit of the New York battery, but produces no effect on the Trenton instrument, as the current does not pass through it. The peg is then transferred from 1-6 to 2-6 (Fig. 127). The circuit of the New York battery is again completed as before, except that it now includes the Trenton instrument, its course being from the line wire L through the instruments as usual, and through the peg 2-6 to the wire g, and finally to the ground. Trenton now becomes a terminal station, working with New York by means of the battery at the latter place.

This result demonstrates to the operator at Trenton that the disconnection is between that place and Philadelphia. If, on the contrary, the circuit had been completed through the relay when the peg was inserted in 6, it would have shown the break to have been in the opposite direction.

It sometimes occurs that the circuit cannot be completed on either side, and the line appears therefore to be interrupted in both directions. In that case the trouble is in all probability within the limits

The Wedge Cut-Out. 155

of the way-station itself, usually in the instrument connections, which should be carefully inspected.

279. **Reporting Result of Test.**—If the operator at a way-station finds by the above test that the line is interrupted in a particular direction, it is his duty to report the fact at once to the terminal station in the opposite direction, from which he should receive instructions in regard to his proper method of procedure, so that the uninjured portion of the line may be operated until the difficulty is removed.

280. **The Wedge Cut-Out.**—This device is often used at way-stations instead of the switch last described. It is termed a *plug*, or more properly a *wedge cut-out*. The instrument wires, i^1 and i^2 of Fig. 122, are connected to the opposite sides of a *wedge*, as it is technically termed, which is shown, full size, in Fig. 128. It consists of two brass plates insulated from each other by a thin plate of hard rubber, and provided with a handle of the same material. The ends of the line-wire are connected with the two binding screws at the top of the baseboard (Fig. 129). The right-hand binding screw is connected, by a wire under the baseboard, with an elastic brass strip. The upper end of this strip is rigidly attached to the board, while the lower end is armed with a brass pin, which, by the elasticity of the strip, is pressed firmly against a second pin, also screwed to the board. The stationary pin is attached by means of a wire to the other binding screw, and is thus placed in connection with the line-wire. This device is termed the *spring-jack*. When the wedge, carrying the flexible instrument wires, is inserted between the two pins, it separates them, thereby breaking the circuit of the main

FIG. 128. Switch Wedge or Plug.

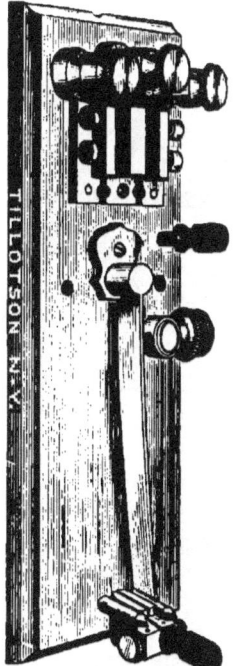

FIG. 129. Spring-Jack and Wedge.

line, but simultaneously opening a new path for the current through the two parts of the wedge and the instruments. Thus the latter may be inserted into or withdrawn without interrupting the main circuit, by a single instantaneous movement. For many places this is an exceedingly simple and effective arrangement. It is sometimes used for large stations, in combination with the peg-switch, as will be hereafter shown (286). The wedge cut-out is usually provided with a lightning arrester and ground-wire connection, arranged with pegs, as in Fig. 129, in much the same manner as the switch in Fig. 122.

281. **Multiple-Wire Switchboards.**—It very often happens that different lines traversing the country in the same or in different directions pass through the same way-station. The necessities of the service sometimes require that there should be apparatus provided for each line, but more frequently a smaller number of instruments is sufficient, provided means are furnished by which any one of them can be inserted into the circuit of any line at pleasure.

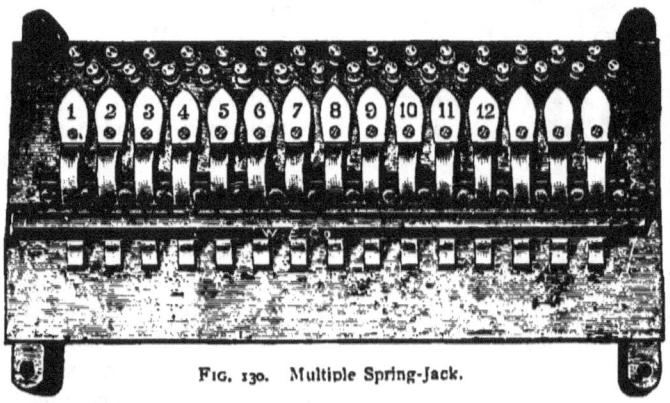

FIG. 130. Multiple Spring-Jack.

282. **Multiple Spring-Jack.**—The simplest way of providing for this is to make use of a number of spring-jacks (Fig. 130) corresponding to the number of line wires, and placed side by side upon the same baseboard. The wires for each separate set of instruments terminate in a wedge, and by this means any instrument may be placed in connection with any required line-wire at a moment's notice, simply by inserting its wedge into the corresponding spring-jack.

283. **Universal Switchboard.**—The arrangement last described makes no provision for interchanging the line-wires among themselves—a proceeding which is necessary, for instance, when

Manipulation of the Universal Switchboard. 157

two or more lines running in the same general direction are each interrupted, but at different points. By connecting the uninjured portions of different lines with each other, it is often possible to "patch up" one or more complete circuits for the transaction of business. The wires are interchanged or *cross-connected*, as it is termed, at the different way-stations, in accordance with instructions given by the official in charge of the circuits at the terminal station. In order to conveniently accomplish this, the different wires are brought into a switchboard of sufficient capacity at each of the principal way-stations, and the necessary changes are made upon this without any interference with the lines themselves.

There are many varieties of universal switchboards, but all are constructed upon the same principle. This will be understood by reference to Fig. 131, which rep-

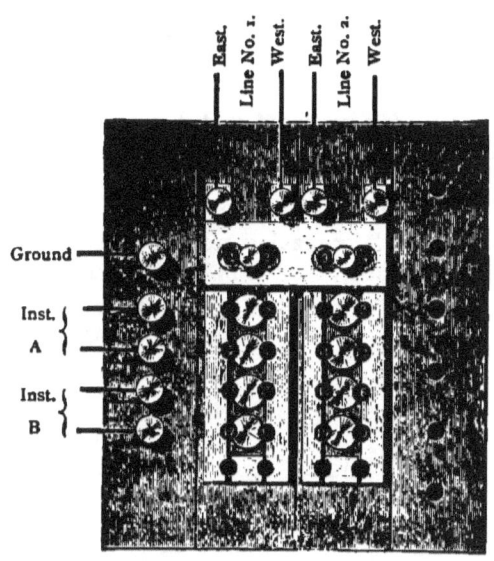

Fig. 131. Two line Universal Switch.

resents a switch designed for a way-station having two line-wires, 1E 1W, and 2E 2W, and two instruments A and B.

284. Manipulation of the Universal Switchboard.—The various changes, other than those which have been explained in connection with the single-wire switch (277), are made as follows:

(1.) *Both lines connected straight, with both instruments in.*—Insert pegs as in Fig. 132, leaving the remaining holes open.

(2.) *Lines cross-connected or interchanged.*—In this case, it is required to connect No. 1 wire west with No. 2 wire east, and No. 1 east with No. 2 west. Insert pegs as in Fig. 133, leaving the other holes open. Both instruments A and B are now included in the circuit. To leave instrument A out of the circuit, change the pegs to the position shown in Fig. 134. Instrument B is cut out by placing the pegs as shown in Fig. 135. In Fig. 136, the wires are interchanged and both instruments cut out, a proceeding which is sometimes necessary when a test is to be made.

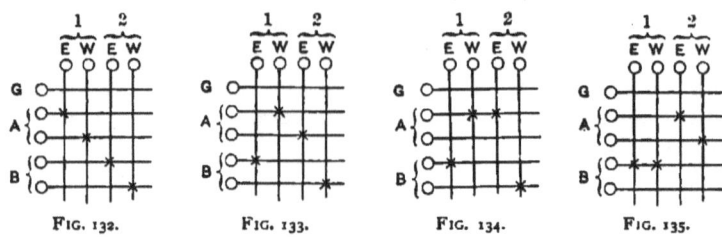

FIG. 132.   FIG. 133.   FIG. 134.   FIG. 135.

(3.) *Lines grounded or put to earth.*—This may be done on either No. 1 or No. 2 wire, east or west, by inserting pegs along the ground wire bar G, as required, as in the single-wire switch (277).

(4.) *Lines looped.*—It is sometimes required to *loop* two wires, as it is termed, for making tests or other purposes. To loop 1 and 2 east, with instrument A in circuit, insert pegs as in Fig. 137; without instrument A, insert pegs as in Fig. 138. Numbers 1 and 2 east may be looped in a corresponding manner.

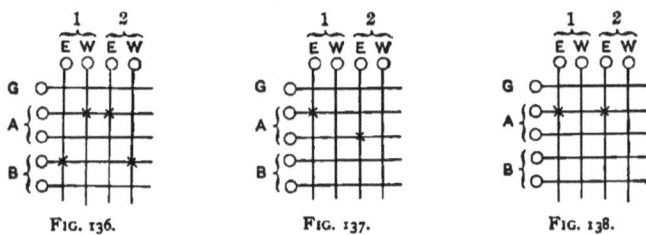

FIG. 136.   FIG. 137.   FIG. 138.

The foregoing explanation will sufficiently illustrate the principle upon which the switch is manipulated. Switches are made to accommodate any number of wires, from 1 to 50 or more.

285. **Arrangement of the Apparatus at the Terminal Station.**—The simplest possible arrangement at a terminal station is similar to that shown in Fig. 122, but it is rare that such a station does not contain more than one line-wire. More than one line entering a terminal station renders it desirable to employ a switch similar in construction to Fig. 131, but with its connections differently arranged, for at a terminal station, provision must be made for connecting and disconnecting the main *batteries* as well as the *instruments*.

286. **Terminal Switchboards.**—Fig. 139 represents a switchboard in a large American terminal station. In this switchboard

Terminal Switchboards. 159

both the peg-switch and the wedge cut-out are employed. In the largest class of terminal stations the switchboard is divided into a number of sections, each section accommodating a certain portion of the lines entering the office. The wires are usually distributed ac-

FIG. 139. Terminal Switchboard for 50 lines.

cording to the geographical location of the region with which they connect. The lines running eastward, for example, are placed in one section of the switch, and those running northward in another, while still another section accommodates the local lines, etc., etc.

The switchboard represented in the figure is provided with 50 vertical bars, to the lower ends of which the line-wires are connected. Between each pair of upright bars is placed a row of metallic disks, to which the battery terminals are connected. All the disks in each separate horizontal row are electrically united at the back by horizontal copper wires, the extreme left-hand disk having a distinguishing number opposite it. The vertical bars are connected at pleasure with the horizontal disks at any required point by the insertion of a peg at the point of intersection. Immediately underneath the lower end of the vertical bars are placed a corresponding number of spring-jacks. Each main wire entering the switch passes first through one of the spring-jacks, and thence to the corresponding bar. Each spring-jack bears an ivory plate, upon which may be engraved the designating number of the circuit to which it is attached.

The baseboard of the switch is of mahogany, cut in strips 2 in. wide by 1 in. thick, separated by a space of $\frac{1}{8}$ in., to prevent injury to the brass-work by shrinkage. Each strip of mahogany supports two vertical bars and one row of disks. The spring-jacks are held in position by stout spiral wire springs attached to the back of the switch. One row of horizontal disks is connected directly with the earth. The lightning arresters are not combined with the switch, as in the smaller stations, but are placed at the point where the wires first enter the building. The instruments seen upon the shelf or counter in front of the switch are used for testing. They are provided with flexible connections and wedges (280), so that they can be thrown into the circuit of any desired line at a moment's notice.

287. Instrument Tables.—In most large stations, the different sets of apparatus are arranged in groups of four, upon tables about 4 by 6 feet, divided by two vertical intersecting screens into four sections, each accommodating a complete set of instruments, sounder, relay, and key. A group of 4 pairs of instrument wires extends from the switchboard to each table, and a second group of 4 pairs of local wires extends from each table to the local batteries. The wires are usually insulated with a double coating of gutta-percha, and are then laid up in cables and bound with tarred tape. It was formerly the practice to group a number of local circuits together, using a common return-wire for the group. Experience has shown that this arrangement is objectionable, and that it is better to keep all the circuits, main and local, of each line, distinct from those other lines in the same station.

288. The Lightning Arrester.—This is a term applied to all devices employed in connection with telegraphic apparatus for preventing danger of injury to the instruments and operators by atmospheric electricity, and from the powerful currents employed in electric lighting, which sometimes find their way into telegraphic conductors. Atmospheric electricity, being of enormous potential, will take a short route through a poor conductor, or even through a non-conductor, in preference to a longer one through a better conductor, while the reverse is true in respect to the currents of comparatively great volume and low potential employed in telegraphy.

289. **The Plate Lightning Arrester.**—A common form has a flat brass plate connected with the ground wire. Other plates of brass which rest upon this are electrically separated therefrom by thin sheets of non-conducting material, or by the air. Each of the smaller plates is provided with one or more binding screws for the attachment of the line-wires. Accumulations of atmospheric electricity upon the lines usually break through the insulating material to the ground-plate, and are thus discharged into the ground without injuring the apparatus. Fig. 140 shows one of the most common forms. The ground wire is connected to the upper plate by the binding screw shown at the left, and the line-wires through the smaller transverse plates beneath. The confronting faces of the plates are surfaced with V-shaped grooves, which have been found by experience to facilitate the discharge of lightning. The lightning arrester is often combined with the switchboard, examples of which construction may be seen in Figs. 127 and 131.

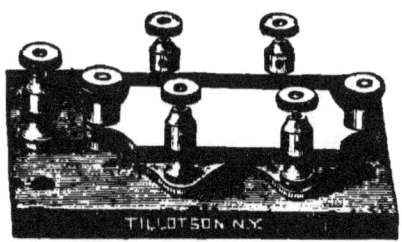

FIG. 140. Plate Lightning Arrester.

290. **The Safety Fuse.**—Another very effectual means of protection consists in interposing 3 or 4 inches of the thinnest copper wire which can be procured (say No. 36) in each circuit between the line and the instrument. An abnormal current, whether arising from atmospheric disturbances or from contact with electric lighting or power circuits, instantly fuses the thin wire, interrupting the circuit, and thus effectually protecting both the operators and the apparatus This device, of course, requires careful attention, inasmuch as it is necessary to replace the fuse-wire after the disturbance has ceased.

291. **Inspection and Care of Arresters.**—Care should be taken to keep lightning arresters, especially those of the plate pat-

tern, free from dirt and moisture. Neglect of this precaution is liable to cause serious interruption of communication. A discharge of atmospheric electricity often forms a permanent connection between the line and ground plates. Hence, arresters should be frequently taken apart and examined, and this should especially be attended to immediately after a thunder-storm.

292. The Repeater.—When the length of a telegraphic circuit exceeds a certain limit, dependent upon the ratio of its insulation to its conductivity resistance, the working margin (220) becomes so small that satisfactory signals cannot be transmitted, even by the aid of increased battery-power. This limit, under the existing conditions of insulation, is much less in wet weather than in fine.

Under such conditions, it was formerly necessary to retransmit all communications at some intermediate station, but this duty is now usually performed by the *repeater*. This is simply an organized apparatus, in which the sounder (or in some cases the relay), receiving the signals through one circuit, opens and closes the circuit of another line, in the manner that a relay opens and closes the local circuit of a sounder (261). The repeater is also used to connect one or more branches with the main line, for the purpose of receiving press-news, etc., simultaneously at widely separated points. Under these conditions the stations in connection may correspond with each other as readily as if all were upon the same circuit. By making use of repeaters it is quite practicable to telegraph direct, when required, between places situated at distance of several thousands of miles apart.

293. Manual and Automatic Repeaters.—The different repeaters which have been devised are almost innumerable. They may, however, be classified as *manual* and *automatic*. The manual repeater is usually employed for temporary purposes, as it requires the constant attendance of an operator to maintain the connections of a switch, in accordance with the direction in which the communication is passing. At repeating stations where a permanent service is maintained, the automatic repeater is employed, which requires no supervision, other than that necessary to insure the apparatus being kept in proper adjustment.

294. The Button Repeater.—A form of manual repeater much used is shown in diagram in Fig. 141. It is known as the *button repeater*. The western main line, after traversing the coil of the relay M^1, passes through the contact-points of armature of sounder S^2 (the movements of which are controlled by relay M^2) and thence to main battery B^1, the opposite pole of which is connected

The Button Repeater.

to ground. In like manner the eastern line traverses the coil of relay M^2 and the contact-points of sounder S^1 and to battery B^2 and ground. It is necessary to provide, in addition, means for " cutting out," or closing the circuit around the breaking-points of each sounder, otherwise the apparatus will be inoperative. For example, suppose the eastern line to be opened by the key of the operator. This allows the armature of relay M^2 to fall off, opening sounder S^2, breaking the circuit of the western main wire at its contact-points. This causes the armature of relay M^2 to fall off, followed by that of sounder S^1, and breaking the circuit of the western line also. The

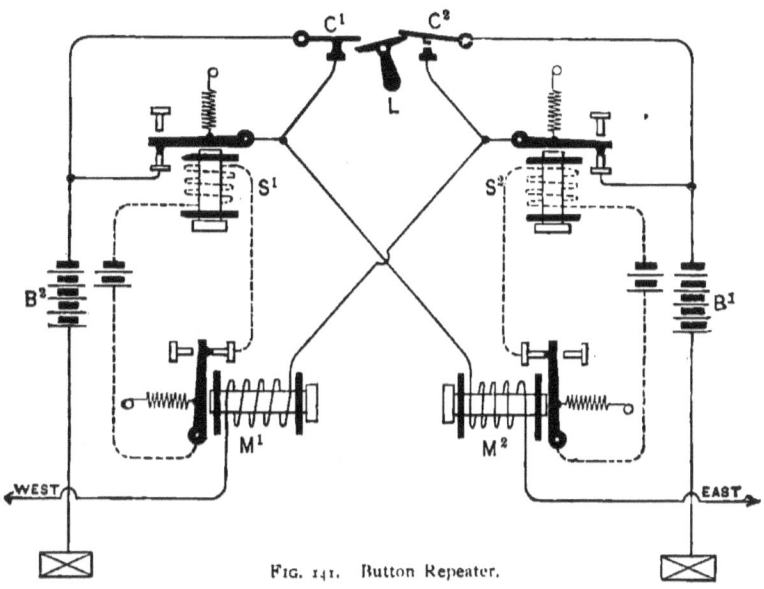

FIG. 141. Button Repeater.

operator of the eastern line cannot now close the circuit, because it is still open in another place, viz., at the contact-points of sounder S^1. The switch shown at L, technically termed the *button*, removes this difficulty, for when it is swung to the right it closes a spring-contact C^1, forming a connection between the contact-points of sounder S^1, enabling the operator of the eastern line to open and close its circuit at pleasure, while his signals are repeated into the western line by the action of the contact-points of sounder S^2. The switch or button shown at L in the diagram, consists of two pairs of contacts C^1 and C^2, normally closed by a spring action, one pair or the other being separated as the handle L is moved to the right or left. If the handle remains in the center, both sets of contacts are closed and the eastern and western lines are entirely independent of each other.

295. **Wood's Repeater.**—Another form of button repeater, known as Wood's, is illustrated in Fig. 142. In addition to the functions performed by the apparatus last described, means are here provided for joining the two lines through in one circuit without repeating. The apparatus of the button is shown in full, with the instruments and batteries, etc., in outline for convenience of explanation. M^1 and M^2 are the eastern and western relays, S^1 and S^2 the eastern and western sounders. The local connections are omitted

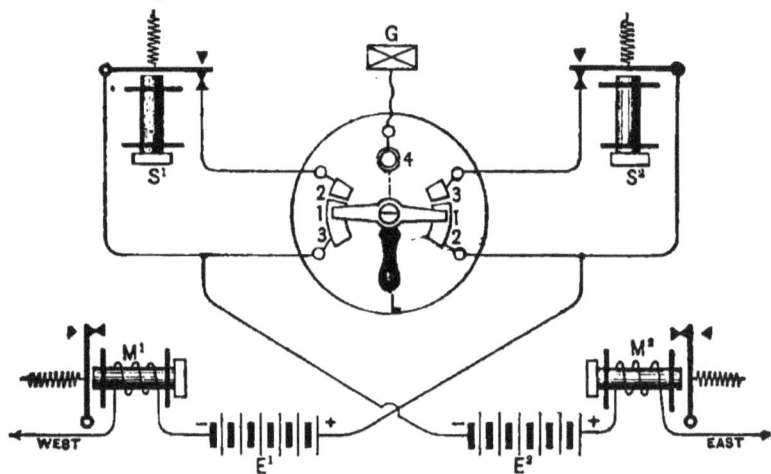

FIG. 142. Wood's Repeater.

o avoid multiplicity of lines, but are run as usual. The eastern and western main batteries at E^2 and E^1 have opposite poles to ground at the repeating station, so that when the lines are connected through, the two batteries will coincide. The following results may be obtained with this apparatus:

(1.) *Two independent circuits.*—The lever L remains in the position shown in the figure marked 1, 1, and the peg at 4 inserted.

(2.) *A through circuit.*—The lever L as before, but the peg at 4 open, interrupting the ground connection between the batteries B and B^1.

(3.) *Two independent circuits arranged for repeating.*—The peg at 4 is inserted. If lever L be placed in the position indicated by the reference figures 2, 2, the eastern sounder repeats into the western

The Milliken Automatic Repeater. 165

circuit. If the lever is shifted to 3, 3, the western sounder repeats into the eastern circuit.

296. Management of Button Repeater.—The duty of an operator in charge of the button repeater is very simple. He has only to keep the relays properly adjusted, and when he hears either sounder fail to work in unison with the other, to instantly reverse the position of the lever L.

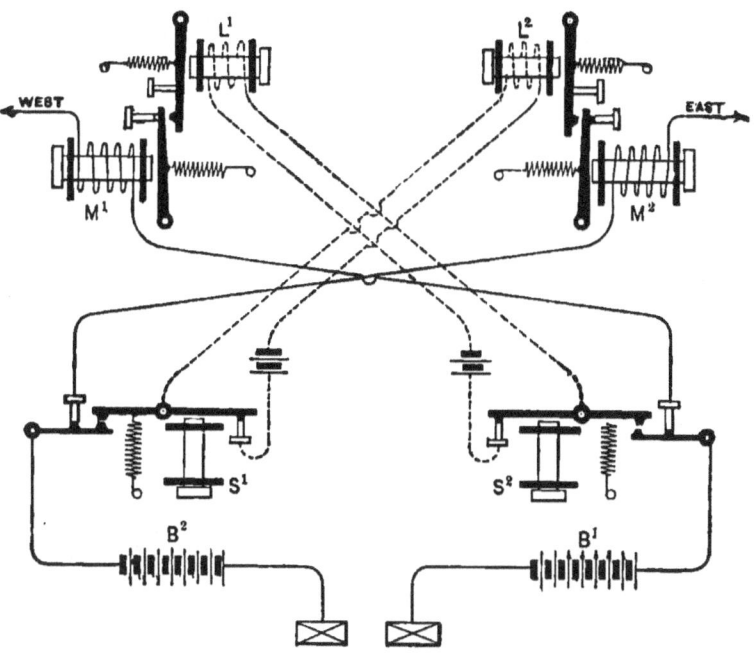

FIG. 143. Milliken's Repeater.

297. The Milliken Automatic Repeater.—This may be considered the standard repeater of the United States, although many others have obtained more or less acceptance. All automatic repeaters embody one essential principle, which is this: The movement of the lever of the relay or sounder on the receiving side of the apparatus brings into action some device for bridging the contact-points of the opposite sounder, before breaking the main circuit on the second line. This is usually effected by some form of spring-contact, although there are different ways in which such a device can be applied to produce the result sought for. The principle of the Milliken repeater is shown in Fig. 143. The main and local circuits are run precisely as in the button repeater (141). The automatic device for closing the opposite circuit is applied to the contact-lever of each of the relays; for example, the relay M^1 has a supplementary local magnet L^1, the armature of which, falling off under the action of its retracting spring, prevents the armature of relay M^1 from likewise falling off, because its spring is adjusted to a stronger tension than the relay-spring. The local magnet L^1 is actuated by a local circuit which is controlled by a contact-point on the lever of the opposite sounder S^2. When the sounder lever falls off, it first breaks the supplementary local circuit, and holds down the armature of the opposite relay, just before the main circuit through that relay is broken. This postponement of the breaking of the main circuit is effected by the spring-contact on the sounder lever. Hence in this repeater, the apparatus on the east side remains quiet while the western line is working and *vice versa*. The Milliken repeater is provided with buttons, not shown in the figure, for cutting out the contact-points, so that the two lines can be worked separately, as in the case of the manual repeaters.

298. **Management of Automatic Repeaters.**—In repeating signals from one circuit to another, the sounder-lever which carries the contact-points has to move a certain distance, after the circuit of the first line is closed, before it can close the circuit of the second line. This occupies a definite time, so that the duration of the current or length of each signal sent forward, is shorter than that received from the transmitting station. A second repeater shortens the signals still more, so that ultimately the signals may fail altogether. This may be partially remedied in practice by the skill of the sending operator, who, in working through a repeater, should transmit his signals more *firmly*, as it is termed, that is, increase the duration of the key contact (374). It is also important that the sounder levers should be permitted the least possible movement compatible with the proper operation of the spring contact-points and with convenience in reading. The armatures of the supplementary local magnets seldom need adjustment if the batteries are kept in good condition. The adjustment of the relays is precisely the same as in ordinary apparatus. The tension of the retracting springs of the sounders, on the other hand, should be very moderate, just enough to raise the armature when released. A repeater works most efficiently when the signals have what is termed a "dragging" sound. When interrupting the sender through a repeater, the receiving operator should first hold his key open for two or three seconds.

Characteristics of the Dynamo-Current. 167

299. The Dynamo-Electric Generator.—In some large telegraphic stations, where the number of lines to be supplied is very great, dynamo-electric machines have been substituted for batteries with highly satisfactory results. A minute description of the different organizations of apparatus or *plants* which have been employed does not properly fall within the scope of the present treatise, but the following explanation may suffice to render the principles of operation comprehensible.

300. Characteristics of the Dynamo-Current.—The theoretical principle of the dynamo has been briefly set forth in a foregoing chapter (80). From the explanation there given, it will be understood that this machine, in its elementary form, produces a series of *waves* or undulations of electromotive force of alternating polarity. Beginning at zero, for example, the *e. m. f.* gradually increases to a positive maximum; then gradually falls to zero, then rises again to maximum negative, then falls again to zero, and so on indefinitely in the manner graphically represented in Fig. 144. Such a condition is obviously wholly unsuited for the telegraphic service, which requires a normally continuous current, of determinate polarity and of approximate uniform strength. The alternate pulsations produced by the revolutions of the armature are therefore *rectified* by means of a device called a *commutator*, the effect of which is to reverse every alternate wave, so as to transform the alternating waves into a series of waves all positive or all negative, as the case may be. If the armature be provided with two coils, placed at right-angles with each other, so that one is in a position of maximum at the same instant the other is in the position of minimum action, and the two effects be combined, the result will be a current which, although continuous, is not steady, as indicated in Fig. 145. By placing a considerable number of separate equidistant coils upon the armature, and superposing their effects, in the manner indicated by the dotted outline at the left, it is practicable to obtain a current which is practically constant, and is found to be perfectly well adapted for telegraphic purposes. For this reason the rotating armature coils of the dynamos employed in telegraphy are divided into a large number of sections, each coming

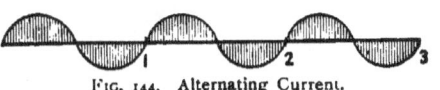
FIG. 144. Alternating Current.

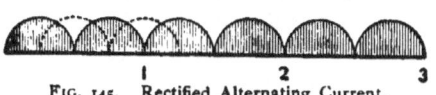
FIG. 145. Rectified Alternating Current.

successively into the position of maximum inductive action as the armature is revolved.

301. **The Electro-Magnetic Field.**—In the theoretical dynamo hereinbefore described (80), the magnetic field in which the armature coils revolve is formed by permanent magnets. In practice, the far more powerful field produced by electro-magnetism is for many reasons preferable. The electro-magnet which maintains the field may be excited by a current traversing a shunt or branch of the armature circuit (140), in which its helix is included, from which circumstance such a machine is termed a *self-exciting* dynamo, and specifically a *shunt-wound* dynamo. Fig. 146 is a theoretical diagram, and Fig. 147 a perspective view of a shunt-wound dynamo such as is used in telegraphy.[2]

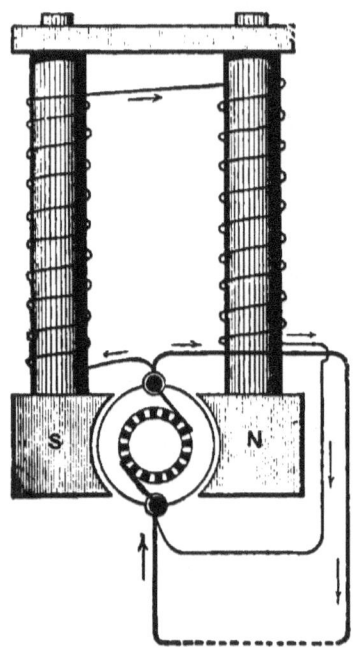

FIG. 146. Circuits of Shunt Dynamo.

302. **The Commutator.**—In the theoretical dynamo (Fig. 26, p. 33), it will be observed that the armature coils terminate in two semi-cylindrical metallic segments carried upon the shaft. Two stationary metallic collectors, termed *brushes*, are made to rub upon the segments at opposite points of the circle as they revolve, the whole apparatus being termed the *commutator*. In Fig. 146 there are a considerable number of segments corresponding to an equal number of armature sections, and two brushes, which form respectively the positive and negative poles of the dynamo-electric generator. The thick black line in Fig. 146 represents the exterior or work circuit. The shunt circuit for exciting the field-magnet is shown by a thin line, the extremities of which are likewise united to the respective terminals or brushes. The direction in which the currents flow in both the main and shunt circuits is denoted by arrows.

---

[2] This particular machine is known as the "Edison No. 2." It is run at a speed of 1200 revolutions per minute, and has a capacity of about 40 ampères. The resistance of the armature is about 0.1 ohm and of the field-magnet coil about 30 ohms.

Characteristics of the Dynamo.

303. Characteristics of the Dynamo.—Each dynamo, when driven at a definite and uniform speed, maintains a practically uniform difference of potential (143) between its terminals or brushes, which is dependent upon the original construction of the machine. This corresponds to the difference which is maintained between the poles of a voltaic element, and is due to the $e.\,m.\,f.$ of the machine, or of the cell, as the case may be. The especial advantages of the

FIG. 147. Perspective View of Shunt-Wound Dynamo.

dynamo over the voltaic battery are: (1) its small internal resistance (131) and consequent capacity to feed a very large number of separate lines without interference, and (2) its economy, both in space occupied and in cost of maintenance, in case the number of wires to be supplied is large.

304. Dynamos in Potential Series.—In a large telegraph station the different lines necessarily vary greatly in their length and resistance, but it is nevertheless requisite that the same quantity of current should be as nearly as possible supplied to each. This renders it necessary that the electromotive forces applied to the respective circuits should differ accordingly. This is effected through the agency of a series of separate dynamos connected together upon the same principle as a series of cells in a battery (132). In the

Western Union telegraph station in New York, for example, there are 5 independent dynamos connected in a series, as indicated in the diagram, Fig. 148. These have potentials as follows: A, 70 volts; B, 70 volts; C, 60 volts; D, 60 volts; E, 65 volts. One terminal of dynamo A is connected to the ground, the wires 1, 2, 3, 4, and 5 are

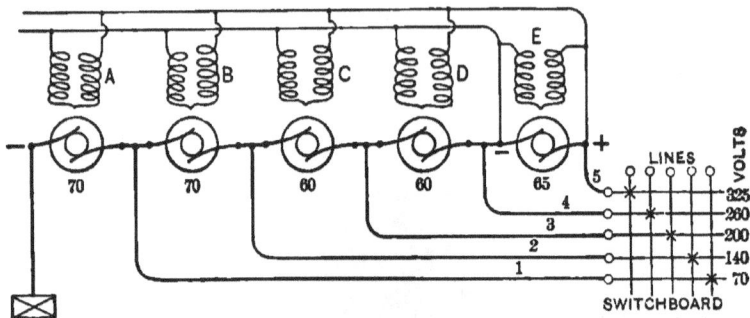

FIG. 148. Dynamos in Potential Series.

led to corresponding horizontal bars on the station switchboard, from the point of connection between each two adjacent dynamos of the series. These bars are respectively termed the first, second, third, fourth, and fifth potentials. The voltage of each potential, and the average resistance of the individual circuits fed therefrom, are as follows:

No. of Potential.	Volts.	Average Resistance per Line.
1	70	3000
2	140	3500
3	200	000
4	260	5000
5	325	5000

As the different lines are each connected to one of the vertical bars on the switch (Fig. 139), of which there may be any number, it will be easily understood how any line may be supplied with the particular potential which it requires, simply by pegging it to the appropriate horizontal bar upon the switchboard.

305. **Positive and Negative Dynamo Series.**—As both positive and negative potentials of the various voltages are required in a large station, two separate series of dynamos are employed, similar to that shown in Fig. 148, one having its positive and the other its negative terminal to the switchboard and line. A third series,

of similar arrangement and capacity, is also provided, which has reversing switches, so that it may be made to send either a positive or negative current to line. This may be used at will as a substitute for either of the two regular series in case of accident. Each series of five machines is driven by a 15 horse-power steam-engine.

306. **Arrangement of the Shunt Coils.**—It will be observed that the last dynamo E in the series (Fig. 148) is necessarily called upon to furnish less current to the lines than the others. The surplus power of this machine is therefore utilized with advantage to furnish current through a shunt or derived circuit, not only for exciting its own field, but the fields of all the other machines in the series. A branch or derived circuit is taken from each terminal of this machine, and the field coils of each of the five machines of the series are connected across from one branch to the other in parallel, as shown by the thin lines in Fig. 148, so that each receives an equal portion of the branch current.

306 *a.* **Capacity of the Dynamo Generator.**—A plant of the capacity of that which has been described can be made to furnish for supplying 1000 lines, and yet is so compact that it may be installed in a small room.

307. **Multiple Telegraphy.**—Within thirty years from the first establishment of the telegraph, the inconveniences arising from the multiplication of wires on the principal commercial routes of the United States proved so serious that it became urgently necessary to adopt measures of relief. The effort to devise an effectual remedy led to the invention of systems of *multiple telegraphy*, in which the same conductor might be used for the transmission and reception of more than one communication at the same time, either in the same or in opposite directions. The most generally useful of these have proved to be those which have been termed the *diplex*, which transmits two messages in the same direction at the same time; the *contraplex*, which transmits two messages in opposite directions at the same time; and the *quadruplex*, which is capable of transmitting two messages in each direction at the same time. The contraplex is more generally known as the *duplex*.

The principle which is common to all these systems is a provision whereby the receiving instrument at the home station, while free to respond to the signals of the key at the distant station, shall not respond to the signals of its associate key.

308. **The Differential Electro-Magnet.**—It has been heretofore explained that the attractive force of the cores of an electro-magnet depends upon the ampère-turns in its coil (176), that is to

say, first, upon the number of turns in the magnetizing helix, and second, upon the strength of current in ampères traversing the wire. It has also been explained that the polarity of the core is determined by the direction of the circulating current (168). It follows from these two considerations, that if two currents of equal quantity are simultaneously made to pass in opposite directions an equal number of times around a magnet core, they will neutralize each other's effect, and no magnetism can be developed in the core. An electromagnet of this kind is termed a *differential magnet*.

309. **Construction of Differential Magnet.** — There are several ways in which a differential magnet may be constructed, all involving essentially the same principle. Two independent wires must be provided for the two opposing currents. These may be wound side by side throughout the whole length of the helix; they may be disposed in concentric independent helices; one helix may be wound on each of the legs of a U magnet, or, what is perhaps most commonly done, the helices may be divided into sections by equidistant non-conducting planes at right-angles to their axes, and the alternate sections connected with the respective circuits.

310. **The Single-Current Duplex.**—The most essential parts of the apparatus of the single-current duplex are the *transmitter*, the

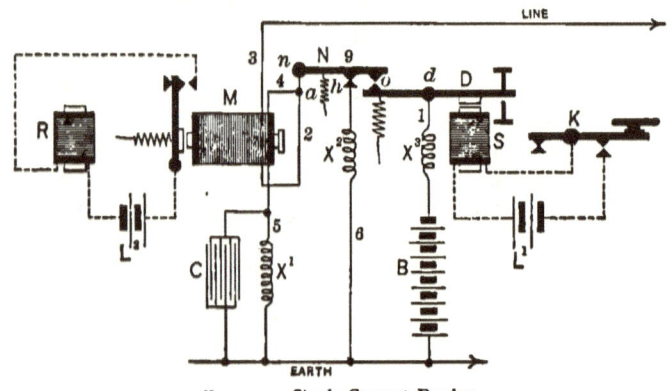

FIG. 149. Single-Current Duplex.

differential relay, the *rheostat*, and the *condenser*. Fig. 149 is a diagram of the apparatus at one of the two terminal stations. The transmitter D is virtually a key, which instead of being actuated directly by the finger of the operator, receives its motion from the armature of an electro-magnet S in the circuit of local battery L^1, which is closed and broken by an ordinary key K. The key-lever

D of the transmitter has a spring contact-lever N pivoted at n and having a contact 9 which normally rests upon a fixed stop h, from which it is lifted by the contact o each time the key K is depressed. The differential magnet is shown at M, and is wound in either of the ways heretofore referred to (309) with two wires, but otherwise does not differ from the ordinary single-wire relay (267). The actual construction of the transmitter is best seen in Fig. 150. The spring-contact here shown differs in form, but not in principle, from that

FIG. 150. Single-Current Transmitter.

outlined in Fig. 149. The spring in this instrument is carried upon the lever, upon a little insulating pedestal, and normally presses upward against a stop affixed to the free end of the lever.

When the armature is attracted, the opposite end of the lever is raised, and the free end of the spring touches the adjustable contact-stop fixed above it in a standard. The object of this device, as will be hereinafter seen, is to close one contact before breaking the other, or in other words, to transfer a current from one branch of the circuit to another without interrupting it.

311. **Circuits of the Single-Current Duplex.**—Tracing the circuit of the line entering the station in Fig. 149, it will be seen to first pass through one wire of the differential relay M, entering by wire 3 and thence going by wire 2 to contact-lever n, and thence through stop 9 and wire 6 to the earth. Hence currents entering from the distant station will actuate relay M and sounder R, pre-

cisely as in the open-circuit system (212). If now, at a time when no current is coming from the distant station, transmitter-lever D be depressed by the closing of key K, the terminal of main battery B will be brought, at the point o, into contact with spring-lever N, and will almost simultaneously lift it from point h. This in effect transfers the in-coming line from the ground-wire 6 to the battery-wire 1, and hence a current will flow through wires 2 and 3 and one coil of the differential magnet m to the line, finding its way to the ground at the distant station. But it will furthermore be observed that a branch leaves the line at the point a, and leads by way of wire 4 through the opposing wire of the differential magnet M, and thence by way of 5 directly to the earth at the home station. Now it is evident that the outgoing current from battery B must divide between the two branches at the point a in the ratio of their respective resistances (134), and if these be equal, the currents must necessarily be equal (140). This equality of resistance is effected by inserting a *rheostat* X^1 (106) of german-silver wire into the branch, so as to make its resistance equal to that of the line. When this has been done, it is evident that the outgoing currents, being equal and opposite in their effects, can produce no magnetism in the relay M, and hence the latter cannot respond in any way to the signals of its associated key K. But this will not in the least interfere with its capacity to respond to the action of currents transmitted by the distant key. In such case one wire of the relay will be traversed by the current due to the conjoint or superposed action of both terminal batteries, while its action is opposed in the other wire of the relay only by the current of the home battery. The incoming current, therefore, produces upon it precisely the same effect as if the current of the home battery were not present.

312. **The Artificial Line.**—The branch from the point a through 4 and 5 to the earth at the home station, is termed the *artificial line*, and the whole problem of contraplex, generally termed duplex telegraphy, consists in simultaneously reproducing in the artificial line, as nearly as may be, all the electrical conditions of the external or working line, and in causing them to act in an opposite sense upon the home relay. There are two conditions in respect to which the actual line is subject to continual variation, viz., as to its *resistance* (115) and its *electrostatic capacity* (147).

313. **Balancing the Resistance.**—The effect of imperfect insulation upon the resistance of the line has already been referred to (242). It is greatest in fine weather and least in wet and foggy weather. The resistance of the artificial line is, however, very easily

Electrostatic Capacity of the Line. 175

balanced; that is, adjusted to correspond with that of the main line by plugging or unplugging, as required, a series of graduated resist-

Fig. 151. [Resistance Coils of Artificial Line.

ance coils of german-silver wire, arranged, as shown in Fig. 151, upon the principle explained in (106).

314. Electrostatic Capacity of the Line.—In order that the process of adjusting the electrostatic balance of the artificial line may be understood, it is necessary first to explain the nature and origin of the corresponding phenomenon as exhibited upon the main line.

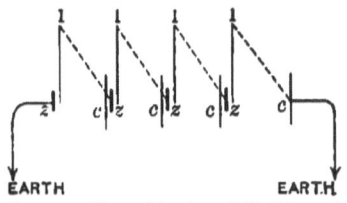

Fig. 152. Recombination of Positive and Negative Electricity.

Whenever electrical generation occurs, there always exists, besides a source of generation at which the normal equilibrium is first disturbed, (1) certain conditions which allow the positive and negative electricity thus separated to recombine, or else (2) certain other conditions by which the electricity generated is accumulated on two surfaces separated by a medium through which it either cannot recombine, or (3) can only recombine less rapidly than the source can generate.[3]

[3] F. C. Webb: *Electrical Accumulation and Conduction* (part 1), p. 2.

The first condition referred to exists in the case of a cell, or series of cells, having the poles joined by a conductor of inappreciable resistance (Fig. 152). The second condition exists in the case of an insulated telegraph line of considerable length, connected to one

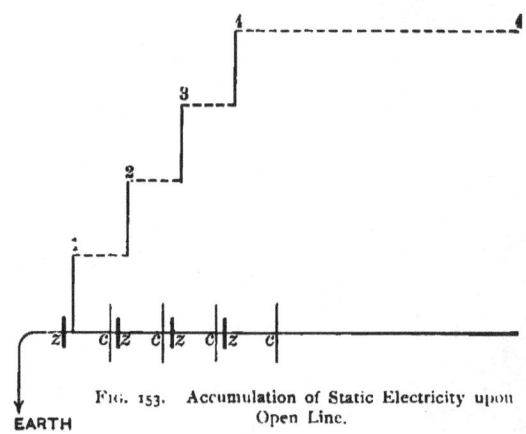

FIG. 153. Accumulation of Static Electricity upon Open Line.

pole of a grounded battery and open at its distant end (Fig. 153). The third condition exists when the line assumed in the last case, instead of being open at the distant end, is there connected to the earth, either directly or through an instrument or other appreciable

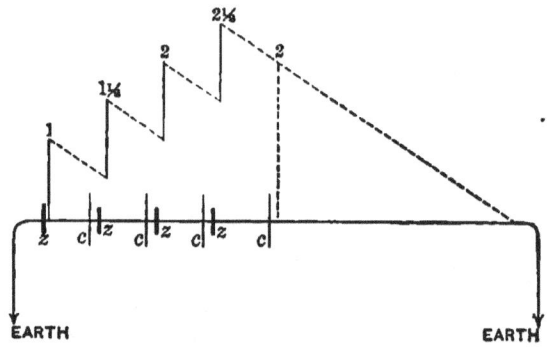

FIG. 154. Accumulation of Electricity upon Line to Ground at Distant End.

resistance. This is the condition of a duplex telegraphic circuit (Fig. 154).[4]

[*] These figures are identical with those which have heretofore been employed in illustration of potential. The area inclosed between the base line and the line of po-

315. **Electrostatic Accumulation upon Insulated Conductor.**—When, therefore, a quantity of electricity flows through a long insulated line, the electricity which constitutes the initial portion of the current, being prevented by the resistance of the circuit from recombining instantaneously, is stored up or *accumulated* upon the surface of the conductor. The quantity thus accumulated depends upon the diameter and length, or, in other words, upon (1) the superficial area of the conductor, (2) its distance from the earth, or other conductors in electrical connection with the earth, and (3) upon the character of the insulating medium which intervenes between it and the earth. Thus, in any long insulated circuit, a certain portion of the current which would otherwise reach the distant station and be available for producing signals, is abstracted and tied up in the form of a *static charge*. If the line be very long and the duration of the current be very short, the static charge may absorb the whole of it, so that no effect will be appreciable at the distant station. As the static charge takes up the initial portion of every current sent, the effect is the same as if its appearance at the distant station were retarded or delayed, and hence the *apparent* velocity of the current is lessened. The initial rush of current into the line, sometimes called the *current of charge*, produces for an instant a much more powerful magnetic effect upon the armature of the relay than does the permanent current which continues to flow after the conductor has been fully charged. The momentary effect thus produced upon the relay is termed by the operators the *kick*. It varies in amount with the electrostatic capacity of the line; the longer the line and the more perfect its insulation, the higher its capacity to receive charge and the more the force manifested by the charging current.

316. **Effect of Currents of Charge and Discharge.**—It will therefore be understood that when one wire of a differential relay is connected, as in the duplex apparatus, with a long insulated line having a considerable electrostatic capacity, while its other and opposing wire is connected with an artificial line principally made up of a rheostat destitute of electrostatic capacity, although the resistances of the two branches may be exactly the same, the initial charg-

tential may be correctly taken to represent the electrostatic charge in each case. A consideration of Fig. 154 will show that when the line is to ground at the distant end, as in duplex telegraphy, 75 per cent. of the aggregate charge of the line is accumulated upon the first half of it, and only 25 per cent. upon the second half. Therefore the currents of charge and discharge are three times as great at the battery end as at the ground end. This is upon the assumption that there is no appreciable loss through imperfect insulation.

ing current in the main line will not be counteracted by a corresponding opposite effect in the artificial line, and hence a momentary *false signal* will be produced upon the home relay. So also, at the termination of the signal, when the line is detached from the battery and connected to earth at the home station, the electrostatic charge accumulated upon the line instantly flows back to the ground through the rear contact of the transmitter, passing through one wire only of the differential relay, and another false signal is produced. These false signals, occurring at the beginning and end of each true signal sent out from the home station, if not eliminated, would mingle with the received signals from the distant station, and utter confusion would be the inevitable result.

317. **The Condenser.**—This difficulty is overcome by connecting to the artificial line a device termed a *condenser*, which consists of a large number of sheets of tin-foil connected with the artificial line, interleaved with an equal number connected with the earth, and separated by sheets of insulating material, usually mica or paraffined linen paper. By arranging these sheets in separate sections, with proper electrical connections, a very large superficial area of tin-foil is exposed to inductive action, the actual extent of which may be varied at pleasure, so that the artificial line may be made to charge and discharge itself at the same instant, and to the same extent, as the main or actual line. The disturbing effects of the static charge and discharge upon the differential relay may thus be wholly eliminated.[5] The condenser is inclosed in a wooden box and provided with a peg switch, as shown in Fig. 155, so that its electrostatic capacity may be varied as required.

The manner in which the condenser is connected to the artificial line and to the earth is indicated at C in Fig. 149. At the common point 5, which is the terminal of the opposing or equating coil of the differential relay, are attached the rheostat X^1 (the resistance of which is maintained equal to that of the main line), and the condenser C (the electrostatic capacity of which is also kept equal to that of the main line), which, by their joint action when properly adjusted, insure a perfect working balance to the apparatus under all conditions.

[5] The credit of the idea of giving to the artificial line of the contraplex apparatus an electrostatic capacity corresponding to that of the main line, is due to Joseph B. Stearns, who successfully applied it on the lines of the Western Union Telegraph Company leading out of New York, in 1872. By this admirable application of a scientific principle, in a manner no less ingenious than simple, it is not too much to affirm that the commercial value of the aggregate telegraphic property of the world was more than doubled at a single stroke.

The Ground and Spark Coils. 179

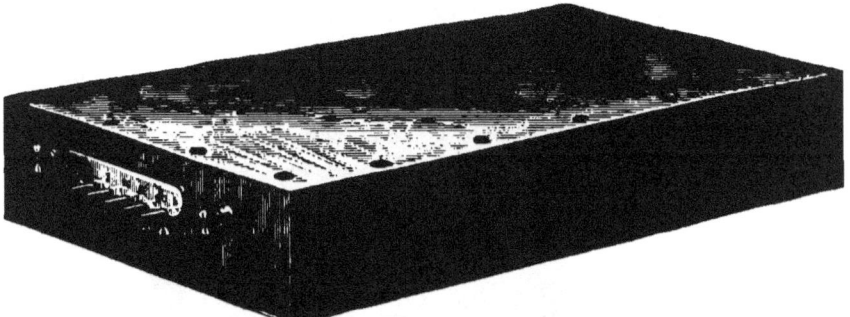

FIG. 155. Adjustable Condenser of Artificial Line.

318. The Ground and Spark Coils.—It is necessary that the apparatus at the home station should always present an equal resistance to the currents coming from the distant station, so as not to overthrow the balance of the distant relay. This is effected by means of small rheostats X^1 and X^2 (Fig. 149), termed respectively the *ground-coil* and the *spark-coil*, the resistance of the former being made equal to that of the latter, plus the internal resistance of the battery, and that of the latter being sufficient to prevent the polarization of the battery when momentarily short-circuited at the transmitter.

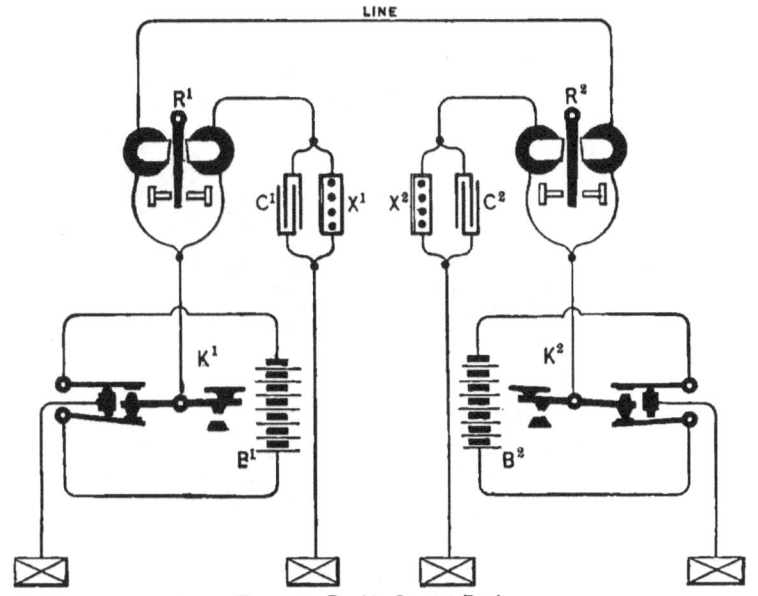

FIG. 156. Double-Current Duplex.

180   *Equipment of American Telegraph Lines.*

319. The Double-Current Duplex.—The principle of this apparatus will be understood by reference to Fig. 156. *Double current* or *reversing keys* are employed at each end of the line, each of which, when depressed, reverses the poles of its associated main battery without interrupting the circuit, and of course without chang-

FIG. 157.   Polar Relay

ing the total resistance.   Polar relays are also used, the peculiarity of which consists in the employment of polarized armatures, the construction and use of which have been explained in (200). The actual construction of the polar relay is shown in Fig. 157.

The permanent magnetism of a polarized armature causes it to remain, by its own attraction, indifferently in contact with either pole of the electro-magnet, when no current is passing through its

The Double-Current Duplex. 181

coils. If, then, we assume this electro-magnet to be differentially wound, and the armature normally at rest upon the rear contact so as to leave the local circuit open, it is obvious that whether the current sent out from the associated key be positive or negative, it cannot in either case, owing to the differential action, develop any magnetism in the cores, nor alter the position of the armature. The

FIG. 158. Pole-Changing Transmitter.

differential winding therefore produces the same effect with the polar as with the neutral relay. But at the distant station, the currents received over the line traverse one wire of the relay only, and hence the polarity of the core at that station is reversed each time the sending key is depressed or raised. When the key is down, a positive current passes, and the armature closes the local circuit; when it is up, a negative current passes, and the local circuit is broken. In other words, the signals are produced by *changing the polarity* of the current, and not by changing its strength from zero to maximum, as in the single-current system. The actual construction of the transmitter employed is shown in Fig. 158.

320. The Quadruplex.—This apparatus may be regarded as a combination of the single-current and double-current duplex systems, adapted to be operated simultaneously in the same circuit. It has been explained that the polar relay in the double-current duplex (319) is actuated solely by changes in polarity, irrespective of strength, and in the single-current duplex (310) solely by changes in strength of current, irrespective of polarity. If then we place both a polar and a neutral relay in series in the same circuit, as shown in Fig. 159, it is evident that we may produce signals by moving the armature of the polar relay to-and-fro by means of alternate positive

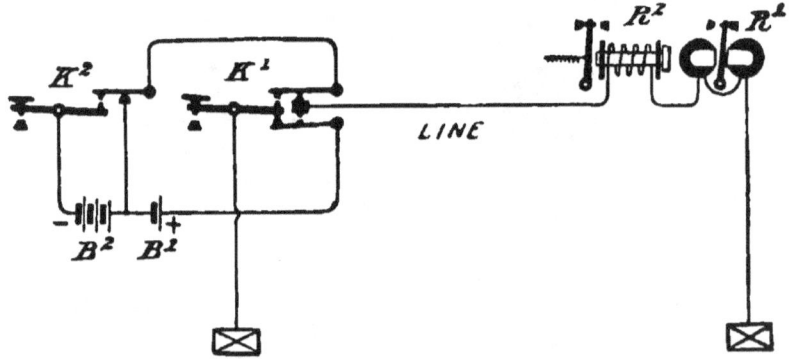

FIG. 159. The Diplex—the Basis of the Quadruplex.

and negative currents, and in case these are not strong enough to affect the armature of the neutral relay, no signals will be indicated thereby. So also, if we maintain a constant polarity, and merely open and close the circuit, we shall produce signals upon the neutral but not upon the polar relay.

321. Principle of the Diplex.—Fig. 159 is a diagram showing a line having at one terminal station two keys, one single-current and the other double-current or reversing. The battery is also in two sections, one section B^2 having three times as many cells and therefore three times as much $e.\ m.\ f.$ as the other section B^1. By tracing the connections, it will be observed that both sections of the battery are included in a *loop*, the terminals of which are reversed by the depression of the key K^1, but that the greater section B^2 of the battery only comes into action when the other key K^2 is depressed.

Operation of the Diplex. 183

Thus it is obvious that the $e.\ m.\ f.$ of the current going to line will be four times as great when key K^1 is depressed as when it is at rest, and that the key K^2 when depressed serves to reverse whatever $e.\ m.\ f.$ may be in circuit at the moment, whether of both sections or only the smaller section of the battery. In brief, the key K^1 controls the *polarity* of the outgoing current regardless of its $e.\ m.\ f.$, while key K^2 controls the $e.\ m.\ f.$ of the outgoing current regardless of its polarity.

322. Turning now to the receiving apparatus, we have in series, at the receiving end of the same circuit, a polar relay R^1 and an ordinary or neutral relay R^2. If the armature-spring of the neutral relay R^2 be adjusted to such a tension that it cannot respond to the comparatively weak current of the battery B^1, when unassisted by the battery B^2, it is obvious that signals may be sent by reversing the smaller battery section by means of the key K^1, which will actuate the polar relay R^1, but will produce no effect whatever upon the relay R^2. On the other hand, signals may be sent by the key K^2, each depression of which throws upon the line the additional $e.\ m.\ f.$ of the greater battery section B^2. The additional strength of current which now flows will actuate the neutral relay R^2, but will produce no effect upon the polar relay other than to increase the pressure of its armature against whichever stop it may chance to be in contact with. But the polarized armature, on the other hand, will instantly respond to each reversal, whether of the smaller or the larger current.

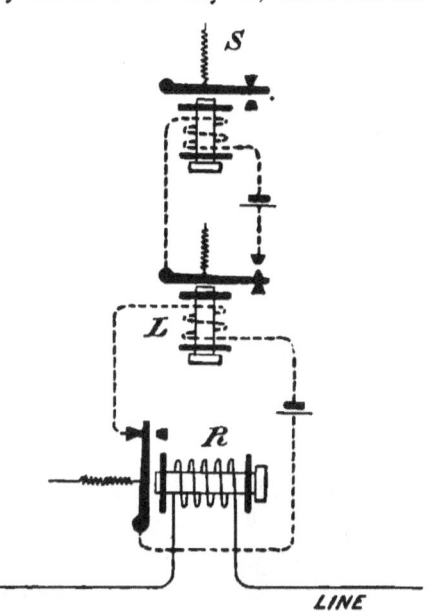

FIG. 160. Receiving Sounder of Diplex.

323. **Operation of the Diplex.**—Suppose a signal is being sent by the depression of key K^2; both sections of the battery are in circuit on the line, causing the armature of the neutral relay R^2 to be attracted. If now another signal be sent by the depression

of the key K^1, the full strength of the current traversing the neutral relay R^2 will be *reversed*. It is obvious that during this operation, no matter how instantaneously the reversal may be effected, there must be an interval during which no magnetism is manifested. The actual result of this is found to be that the neutral relay lets go its armature for an instant, and the spring begins to pull it away, but it scarcely has time to move before the opposite magnetism seizes upon it and restores it to its original position. This, if not guarded against, causes a slight break in the signal, known as the *clip*, which may nevertheless be eliminated by the aid of special devices.

One of the most efficient of these devices is that of an intermediate local relay interposed between the neutral main-line relay and its associated sounder in the manner indicated in Fig. 160. When the armature of the neutral relay R falls off, the sounder S is not affected until it reaches its rear contact-point, when it closes the circuit of the local relay L, and the latter, also by its rear contact, breaks the second local circuit. When the main-line is closed the reverse action takes place. Thus the sounder can only be affected by a full opening of the main circuit, which shall continue long enough to permit the relay armature to reach in rear contact. A

FIG. 161. Short-core Neutral Relay for Quadruplex.

neutral relay having very short cores, as in Fig. 161, is for this reason advantageous (195).

324. The Diplex and Contraplex Combined.—Having an apparatus of this kind, capable of transmitting two sets of signals in the same direction at the same time without interference with each other, it is not difficult to understand that by applying a differential winding to both relays, polar and neutral, and by including both in the circuit of the main and artificial lines, precisely as in the case of the single-current and polar duplexes (310, 319), it becomes perfectly practicable to transmit two sets of signals upon a line in each direction at the same time, and this is in fact precisely what is done in the case of the quadruplex.[6]

325. Quadruplex worked by Dynamo-Currents. — The quadruplex apparatus at the larger stations in the United States is now frequently operated by dynamo-currents, and it is probable that this method will in time become practically universal.[7] The organization of the apparatus has been slightly modified from that illustrated in Fig. 159, to better adapt it to the conditions under which it is required to work. The principle will be understood by reference to Fig. 162. D^1 and D^2 represent two independent series of dynamos, such as hereinbefore described (304), one having its positive and the other its negative pole to the line. K^1 is the pole-changing transmitter and K^2 the single-current transmitter, which, for simplicity, are shown in the diagram as keys, but which are in practice operated by electro-magnets, local batteries, and independent keys, as indicated in Fig. 149. When the apparatus is at rest, the current from the negative dynamo D^2 traverses a resistance coil of say 600 ohms (which is inserted to avoid danger of injury to the instruments in case of an accidental short circuit) to the rear contact of the pole-changing key K^1; thence through wire 1 (in which is included a rheostat of say 1200 ohms) to the point 2, where it divides into three portions; the first portion going to the line and distant station, the second through the artificial line, including rheostat X, to the earth, and the third through the wire 3, the normally closed rear contact of the single-current key K^2, and a rheostat of say 900

---

[6] The method of diplex transmission here described, which forms the basis of the commercial quadruplex system, was invented in 1873 by Thomas A. Edison (see U. S. patent No. 162,633, Apr. 27, 1875). He also devised the apparatus described in (323), to overcome the principal obstacle in applying the method in quadruplex transmission.

[7] The first successful application of the dynamo machine as a substitute for the voltaic battery in commercial telegraphy was made in 1879 by Stephen D. Field of San Francisco. (See his U. S. patents, Nos. 223,845, Jan. 27, 1880, and 243,698, July 5, 1881.) Detailed descriptions of some of the more important dynamo plants have been published as follows: Western Union, New York, *Operator and Electrical World*, xiv. 225; same plant improved, W. MAVER, Jr., in *Electrical World*, xi. 67, 79; W. U. plant, Pittsburgh, W. MAVER, Jr., *ibid*, xii. 195; W. U. plant, Chicago, *ibid*, xv. 173; Postal Tel. Cable Co., N. Y., *ibid*, xii. 65; Postal T. C. Co., Boston, *ibid*, xvi. 313. The plant of fifteen dynamos in the Western Union N. Y. central station does the work of more than 30,000 cells of gravity battery.

186 *Equipment of American Telegraph Lines.*

ohms, to the earth. If for example, therefore, we assume the resistance of the main and artificial line to be 3600 ohms each, it follows from the law of distribution of currents in branch circuits (140), that

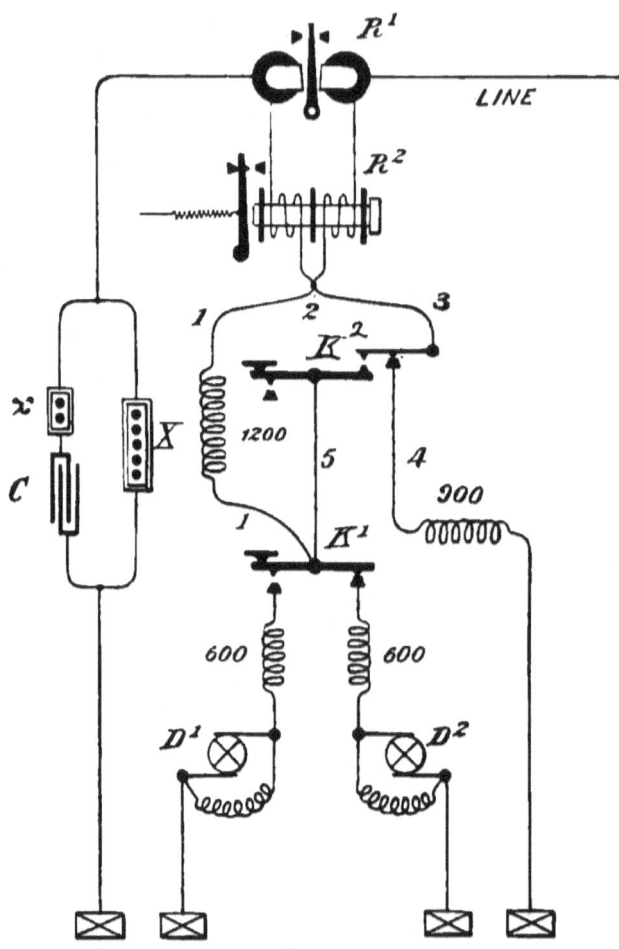

FIG. 162. Quadruplex Operated by Dynamo Currents.

two-thirds the current will return to earth through wires 3 and 4, one-sixth will go to the main line, and one-sixth to the artificial line.

326. Distribution of Currents in Quadruplex Apparatus. —If now the key K^2 be depressed in order to send a signal, a direct connection will be formed between key K^1 and the point 2 through wires 5 and 3, shunting the 1200-ohm coil in wire 1. At the same time the wire 4 will be opened, and the whole current will divide at the point 2, half going to the main line and half to the artificial line. It follows, therefore, that with the several resistances in the ratios shown, the current sent to line by the key K^1 when key K^2 is depressed will be three times as strong as when the latter is raised, and this will be equally true whether the current sent by key K^1 be positive or negative.

327. A computation of the effects of the several resistances will also show that when an arriving current reaches the point 2, the resistance which it has to encounter in passing thence to the ground is the same, whether the key K^2 be depressed or raised. When the key is depressed, the resistance is only that of one or the other of the 600-ohm coils between the key K^1 and the dynamos; when raised, it is the joint resistance of one coil of 600, plus the coil of 1200 (a total of 1800), in one branch, and the coil of 900 in the other branch, the joint resistance of the two being 600, the same as in the first instance. The relays R^1 and R^2 at each station, being both differential, are not affected by outgoing currents, whatever may be the strength or the polarity of such currents.

328. **Practical Management of the Quadruplex.**—Skill in the management of an apparatus of so much complexity as the quadruplex can only be acquired by experience and careful study. Only a few hints can be given here.[8] As a preliminary to these suggestions, an explanation of certain technical terms which have come into use with the apparatus is necessary.

The " No. 1 side " of the apparatus comprises the pole-changing transmitter, the polar relay, and their attachments.

The " No. 2 side " of the apparatus comprises the single-current key, the neutral relay, and their attachments.

The *tap-wire* is the intermediate wire which divides the battery into two unequal portions, usually termed respectively *the long end* and *the short end*. (See Fig. 159.)

The resistance which is inserted to compensate for the internal resistance of the battery is called the *ground-coil*. A resistance x, Fig. 162, is also placed between the condenser and the differential

---

[8] A great amount of information of value respecting the history, theory, and practice of quadruplex telegraphy may be found in a series of articles by WILLIAM MAVER, Jr., in *Electrical World*, xi. 254, 266, 280 ; and in a subsequent discussion by FRANCIS W. JONES, *ibid*, xi. 290, 330, xii. 276 ; W. MAVER, Jr., *ibid*, xi. 305, xii. 231 ; CLARENCE L. HEALY, *ibid*, xi. 292 ; THOMAS HENNING, *ibid*, xi. 330 ; H. W. PLUM, *ibid*, xi. 316. See also F. W. JONES: The Quadruplex, *Journal Am. Electrical Soc.*, i. 16 ; F. L. POPE : Quadruplex Telegraphy, *The Telegrapher*, xi. 271.

188 Equipment of American Telegraph Lines.

relays on long circuits, to retard the time occupied in the charge and discharge of the condenser, in correspondence with that of the line.

The proportion between the long and the short end of the battery varies in practice with the length of the line. On lines of 100 miles or less the division is usually equal, or, as it is termed, 2 to 1 ; on a line 350 or 400 miles, it may be with advantage as much as 4 to 1.

329. **Adjustment of the Apparatus.**—The following method of procedure has been recommended by experienced operators, though it is proper to say that some difference of opinion exists in reference to the minor details of adjustment.

First. Instruct distant station to "ground." He will then put the line to ground through his battery-compensation resistance or ground-coil. Both stations should assure themselves that the resistance of the ground-coil is equal to that of the battery.

Second. The line being to ground at both ends, proceed to *centre the armature* of the polar relay. When centred, it should remain indifferently in either an open or closed position of the local circuit as placed by the finger.

Third. Switch in the home battery, and vary the rheostat resistance in the artificial line until the polar relay can be again centred. If disturbing effects from foreign currents are felt, it may not be possible to do this accurately. In such case, approximate it as nearly as possible.

Fourth. Instruct the distant station to switch in his battery. This may assist in adjusting the polar armature.

Fifth. Instruct distant station to close both keys, thus sending full current to you. Close your No. 2 key ; send dots on your No. 1 pole-changer, and alter the capacity of your condenser until its effects on the home polar relay are eliminated. This condition is termed the *electrostatic balance.*

Sixth. After both stations have thus balanced, test the correctness of the adjustments as follows :

Instruct distant station to send dots on No. 1 and words on No. 2. While this is being done, alternately open and close both keys at the home station. If both sets of signals from distant station come distinctly under all circumstances, the balance is obviously correct. The same test should be repeated by the distant station, in order to ensure an accurate working adjustment.

In the above test, if the sending on No. 2 side should fail to come well, instruct distant station to hold No. 1 key open for a few seconds, and then closed the same length of time. If the signals come imperfectly in both cases, it indicates that the contact-points of the distant pole-changer require cleaning. A very fine flat file is the proper tool to use for this purpose.

If the dots on No. 1 fail to come well at the same time with the writing on No. 2, instruct distant station to alternately open and close No. 2 key at intervals of a few seconds; the trouble may usually be traced to defective contacts upon the single-current transmitter, provided the balance has been properly attended to.

It should not be forgotten that a change of weather which is sufficient to affect the insulation of the line, may necessitate a readjustment, to a greater or less extent, of both the rheostat and condenser balance of the quadruplex. Both the line resistance and the electrostatic capacity are diminished by a defective state of insulation.

The difficulties which may arise in the operation of a quadruplex apparatus are of such various character that it would be quite impossible to enumerate them in detail. Those which have been referred to are among those most liable to occur under ordinary circumstances.

329*a*. **Repeaters for Multiple Telegraph Systems.**—Notwithstanding the apparent complexity of the duplex and quadruplex apparatus, the arrangement of repeaters in connection with them is a very simple matter. It is only necessary to include the electro-magnet which works the transmitter of one line of communication in the same local circuit with the receiving apparatus of another line of communication. This facility of adaptation of repeating devices gives great flexibility to the system, and enables it to be employed for special service in a great variety of ways. Thus, for example, a single wire might be used as a duplex between New York and Boston, New York and Hartford, and Hartford and Boston. The local circuits of the transmitter and of the receiver in a main office may be extended through several branch offices in the same city, and thus all these branch offices may exchange messages directly, either with a distant main station, or with any one of a similar group of branch offices in the distant city. The limits of the present work will not permit a detailed description of these various applications of the system, with their numerous modifications, but the general principle will be readily comprehended by those who have made themselves familiar with the apparatus on which they are based.

CHAPTER IX.

TESTING TELEGRAPHIC LINES.

330. **Object of the Tests.**—Telegraphic lines, from their exposed situation, are peculiarly subject to interferences and interruptions from various causes, and hence one of the most important duties of an operator is to familiarize himself with the nature of these disturbances, so that their location may be quickly determined and the proper measures taken for their removal. This is effected by an experimental investigation, technically termed *testing*. Another object in testing is to examine the electrical condition of the wires at stated intervals, and thus detect incipient faults before they become serious enough to cause interruption of the service.

331. **Faults and Interruptions.**—The principal sources of interruption may be classified as follows:

(*a*) *Disconnection or Break.*—The continuity of the circuit is interrupted. A break may give rise to three different conditions: (1) When neither of the broken ends is in electrical connection with the earth; in this case the circuit is wholly interrupted so that no current can pass; (2) when one end is in connection with the earth; in this case there is no current on the portion of the line which is disconnected from the earth, but more or less on the other portion; and (3) when both ends are in connection with the earth, in which case there will be more or less current on both sections of the line.

(*b*) *Partial Disconnection, or Resistance.*—This fault may arise from unsoldered and rusty joints in the line-wire; from loose connections in the offices, or about the instruments, switches, and batteries; or from a defective or insufficient contact between the earth-plate and the earth.

(*c*) *Escape.*—Leakage from the line to the ground arising from defective insulation generally, or specifically from the line getting into contact with trees, wet buildings, etc. When an escape is so serious that it is impossible to work past it, it is called a *dead ground*.

(*d*) *Cross.*—This term is used to denote a leakage or escape of current from one wire into another. An absolute contact between

Testing for Disconnection. 191

two or more wires, so that current passes freely from one to the other, is termed a *metallic cross*. Sometimes parallel wires on the same supports swing to and fro in the wind, occasionally touching each other, and causing an intermittent disturbance termed a *swinging-cross*. When only a portion of the current passes from one wire to the other, through defective insulation, wet cross-arms or the like, the effect is termed *cross-fire*, or sometimes *weather-cross* (248).

A defect in a ground wire or plate which serves as a common terminal for two or more lines, produces an effect similar to that of a metallic cross. This difficulty not infrequently arises from the removal of a meter in the line of a gas-pipe which is used as a ground connection for the wires (210).

332. Testing for Disconnection.—When the circuit of a line operated on the American or closed-circuit system (214) is totally

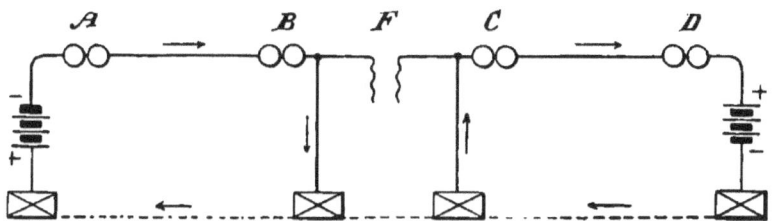

FIG. 163. Test for Disconnection.

interrupted, the armature of every relay in the circuit will fall off. In such case, the operator at each way-station should immediately proceed to test the line by connecting his ground to the line, first on one side, and then, if necessary, on the other, as has been explained in a previous chapter (278). If either connection closes the line, the interruption is on that side, for the circuit of the opposite terminal battery is completed through the ground in place of the interrupted wire. If the ground-wire gives no current on either side, it is most probable that the trouble is in the testing-station, though it may be that the ground connection is defective. Each operator should first assure himself by a careful examination that the fault is not in or about his own apparatus.[1] Having ascertained the direction in which the difficulty lies, he should at once report the facts to the terminal station at the opposite end of the line.

[1] An easy and expeditious way of doing this is to open the key, and then slightly moisten one finger of each hand and touch lightly the binding-screws by which the line-wires enter the switch. If the break is within the office a current will be perceived by the touch.

Fig. 163 represents a line with four stations, *A*, *B*, *C*, and *D*. If, for example, the line be interrupted by a break at *F*, two operative circuits may be formed by putting on the ground-wires at *B* and *C*, as shown in the figure. *A* can work with *B*, and *C* with *D*, notwithstanding the break between *B* and *C*.

333. Disconnection is usually caused either by the breaking of the line-wire or else by the careless leaving open of the switch of a key. Other less frequent causes are wires loose in their binding-screws (a defect peculiarly liable to occur in railway-station offices, on account of the continual vibration caused by the passage of trains), defective switches, and breakage of the fine copper wire in and about the relay. Sometimes the latter is burned in two just where it enters the helix, by the action of lightning.

334. **Testing for Partial Disconnection.**—It is somewhat difficult to locate this fault by testing merely with a key and relay. It is liable to be of an intermittent character, which by no means tends to lessen the difficulty. In case of this or any intermittent fault, the best plan is to cross-connect, where it can be done, by interchanging the defective line with a good one at the terminal and

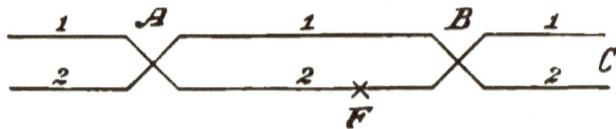

FIG. 164. Test for Intermittent Fault.

also at some other station, as in Fig. 164. If, for example, the fault is at *F* on No. 2 wire; by cross-connecting at *A* and also at *B*, as shown, the fault will shift to No. 1, showing it to be between the two points where the wires were to have been interchanged. If it were beyond *B* it would have remained on No. 2 circuit. In the latter case, put the wires straight again at *B*, and cross-connect at *C* and so on, station by station, until the fault shifts to No. 1, which proves it to be between the two last stations.

335. **Testing for Escape.**—Call the stations up in rotation, beginning with the most distant one, and instruct each one to open his key for say ten seconds. When any station beyond the point of escape is open, a weak current will nevertheless pass to line through the home relay, the circuit being partially completed through the ground by the fault. For example, in Fig. 165, assume that terminal station *A* is testing. When the key is open at *C* or *D* a current will

Testing for a Cross. 193

pass to ground through the point of escape, which will disappear when *B* opens his key, showing the fault is between *B* and *C*.

If the escape is so serious as to be in effect a ground, the operator at a way-station can often ascertain in which direction from him the

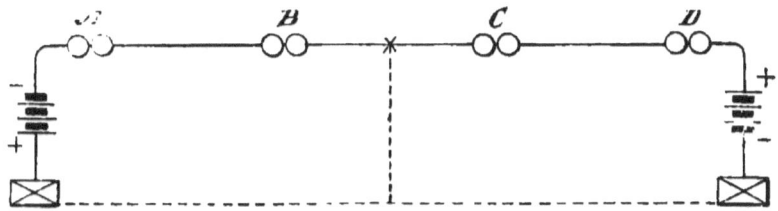

FIG. 165. Test for Escape.

fault is located, from the fact that it cuts off or perceptibly weakens the current from the terminal battery in that direction, when tested with the ground-wire by the aid of the sense of feeling in the finger or tongue.

336. **Testing for a Cross.**—In case a cross is suspected to exist between two wires, as for example No. 1 and No. 2, instruct the most distant station to open No. 1, and send dots on No. 2 wire. Then open No. 2 at your own station, and if the dots sent on No. 2 at the distant station are received on No. 1, the wires obviously must be crossed. Some care is necessary not to be deceived by cross-fire, due merely to imperfect insulation, and not to actual contact between the wires. If the wires are in actual contact, the dots or signals will come nearly as strong on one wire as on the other.

Next, instruct the distant station to leave No. 1 open. Open it at the home station also. No. 2 will now be free from interference, and stations may be communicated with upon it without difficulty. Commencing at the most distant station, call them in succession, and instruct each one in turn to send dots on No. 2. If the dots come on both wires, the cross is between the home station and the sending station, but if upon No. 2 only it is beyond the station sending. Each operator along the line should be instructed, while sending dots on one wire, to open the other wire, if practicable.

337. **Principle of the Cross Test.**—Figures 166 and 167 will explain the principle of this test. A two-wire line is represented having four stations, *A*, *B*, *C*, and *D*. Assume the operator at *A* to be testing for a cross which is located between *B* and *C*. In Fig. 166, No. 1 is open at station *C* and No. 2 is open at station *A*. If *C* sends dots on No. 2 the current will pass over to No. 1 at the

cross, as indicated by the arrows, and the dots will come on the No. 1 instrument at A, showing that the cross is between A and C. In case C is not able to open No. 1, the result will evidently be the same, provided it remains open at D.

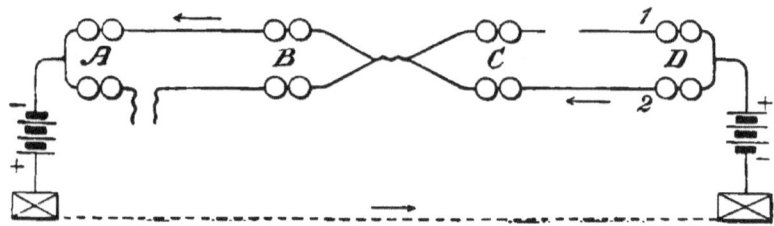

Fig. 166. Test for Cross.

Now, if C closes both wires and B opens No. 1, and writes dots on No. 2, as in Fig. 167, B cannot work when No. 2 is open at A, as both wires are open, one at A and the other at B. With both wires closed at A, the dots which B sends on No. 1 will reach A on No. 2, the current from F going over both the wires to the cross, and from thence to A on No. 2 alone. Thus the cross is definitely located between B and C.

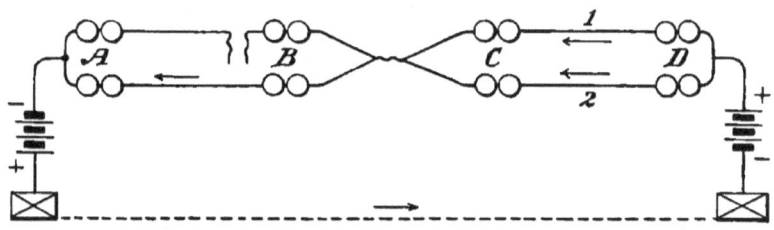

Fig. 167. Test for Cross.

338. A convenient and expeditious method of testing for crosses, in offices where there are a considerable number of wires, is for the operator to station himself at the switch with a test instrument, as shown in Fig. 139. When any station has been instructed to send dots on some particular wire, the testing operator can detect them by placing one finger upon the ground-wire and the other upon the line-wire to be tested, or its corresponding bar upon the switch. In wet weather, however, this method of testing is sometimes attended with much uncertainty, as it is extremely difficult to distinguish by this means between the effect of a metallic cross and those due to the leakage from wire to wire through imperfect insulation.

339. When a cross is found to exist between two lines, the one having the largest number of offices, or for other reasons the most available for business, should be cleared. The remaining wire can then be utilized for a considerable portion of its length, by instructing the stations nearest the cross in each direction to open the main-line wire at the switch, on the side towards the cross; ground the other side, and ground the line in the other direction. This will enable the second line to be utilized in two sections.

340. **Testing by Quantitative Measurement.**—The tests which have been thus far described are such as may be made by the ordinary apparatus employed for the transmission of messages. They serve merely to roughly indicate the nature of the difficulty when it is serious enough to appreciably interfere with correspondence, and to determine between which two neighboring stations it is situated, but for accurate work more refined methods are necessary. Galvanometers and rheostats are the most essential instruments employed for this purpose, and the results are deduced by computation from actual measurements, made upon the principles which have been explained in Chapter IV.

The measurements required are principally of two kinds: measurements of *resistance* and measurements of *quantity*, or, as it is usually termed, current.

341. **The Wheatstone Bridge.**—This apparatus consists of three sets of resistance coils, a galvanometer, a battery, one or more keys, and the necessary connections.[2] Its principal use is to measure resistances, which may be done by its means with great convenience and accuracy, usually from 0.01 ohm up to 1,000,000 ohms. The theoretical arrangement of the bridge is shown in Fig. 168. It consists of four resistances, a, b, d, and x, arranged in a parallelogram, the galvanometer being connected across one transverse diameter, and the battery across the other. When the values of the four resistances are so adjusted in relation to each other that the current from the battery produces no deflection upon the galvanometer, it is certain that these several values must then bear a

---

[2] This ingenious and useful system of electrical measurement was first described by SAMUEL HUNTER CHRISTIE, in *Phil. Trans. R. S.*, 1833, 95–142. Its importance remained unappreciated until attention was directed to it by Professor CHARLES WHEATSTONE, in a lecture before the Royal Society in 1843, entitled "An account of several new Instruments and Processes for determining the Constants of a Voltaic Circuit." *Phil. Trans. R. S.*, cxxxiii. 303–327. Although full credit was accorded to Christie by Wheatstone for his admirable device, electricians have ever since persisted in calling it the *Wheatstone Bridge*, and it seems probable that it will always continue to be known by that name.

determinate ratio to one another. This ratio may be expressed as follows:

As a is to b, so is d to x.

This ratio holds good, entirely irrespective of the magnitude of any of the resistances. In the actual apparatus, therefore, as used in practice, two of the resistances (a and b) are fixed, and the third (d) adjustable, the fourth (x) being that which is to be determined.

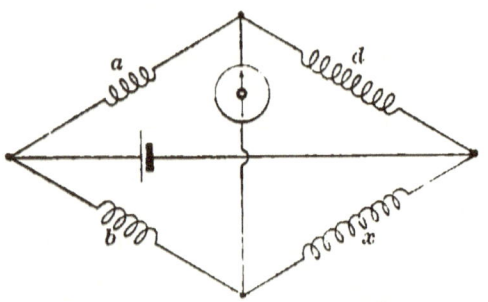

FIG. 168. Principle of Wheatstone Bridge.

342. Best Ratio of Electromotive Forces and Resistances.—In performing the operation of testing, with equal resistances in the branches a and b of the bridge, the most trustworthy determinations are reached by preserving a due relation between the value of the *e. m. f.* employed, the branch resistances a and b, and the unknown resistance x, which is to be measured. Hence when the unknown resistance is

Between 1 and 100 units, a and b should be 10 ohms each; *e. m. f.*, 1 volt.
Between 100 and 1000, a and b should be 100 ohms each; *e. m. f.*, 10 volts
Between 1000 and 10,000, a and b should be 1000 ohms each; *e. m. f.*, 100 volts.

343. Principle of the Wheatstone Bridge.—This may be most readily comprehended by considering that at every point where

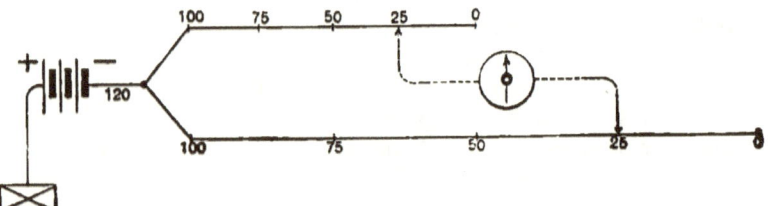

FIG. 169. Fall of Potential in Arms of Bridge.

a circuit divides into two or more branches, the potential of each branch must necessarily be the same. If, at any other point, any two or more of these branches are again joined, the potentials must again be the same. In the bridge, therefore, if we assume the potential at the point where the current first divides to be say 100 volts,

Principle of the Wheatstone Bridge. 197

and at the point where they meet again or are connected to the earth to be o, let each circuit be assumed to be divided into 100 equal spaces, as indicated in Fig. 169. If now a wire be connected across from one of the branch circuits to the other, connecting the point 50

FIG. 170. Wheatstone Bridge Apparatus.

to 50 or 75 to 75, or, as shown in the figure, 25 to 25, or between any other two points whatever having the same potential, no current can flow from one point to the other through the wire, because there exists no difference of potential between its ends; but if, on the other hand, the wire is connected between any two points of different potential, as, for example, from 50 to 25, a current will necessarily flow through it (143), and a galvanometer placed in the wire will be deflected. When, therefore, the needle is not deflected, the proportionality referred to in (341) must always exist between the resistances of the four sides of the bridge.[8]

344. **Actual Construction of the Bridge.**—One of the most useful sets of coils for general purposes is that shown in Fig. 170,

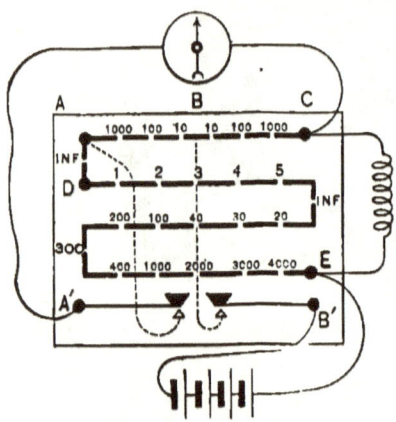

FIG. 171. Diagram of Bridge Connections.

and in outline, with diagram of bridge connections, in Fig. 171. The various resistances are arranged in the manner hereinbefore described (341); and in the diagram, Fig. 171, as well as on the actual box, the respective values of the resistances are denoted by numbers representing ohms. The points marked INF., or "infinite," indicate a total disconnection when the plug is withdrawn.

In some of the more modern sets of apparatus, such as that shown in Fig. 172, the galvanometer, rheostat, branch-coils, contact-keys, and five cells of battery, with the necessary connections, are all put up in a portable mahogany box, with lock and handle. A lifter is provided for raising the needle from its pivot when the apparatus is not in use.

345. **Galvanometer for the Wheatstone Bridge.**—In selecting a galvanometer for any particular purpose, one having a few turns of thick wire, and small resistance, is most suitable for measuring small resistances, while for long circuit or a great resistance of any kind, a galvanometer of many turns of thin wire should be selected. Fig. 173 shows an excellent type of galvanometer for use in the bridge as well as for general purposes. It has an *astatic* system of needles,[4] suspended by a delicate silk fibre, and is fitted

---

[3] The above explanation has been adapted from LATIMER CLARK: *Electrical Measurement*, p. 85. The student desiring to acquaint himself thoroughly with the theory of the bridge may, with advantage, consult also F. JENKIN: *Electricity and Magnetism*, p. 241; H. R. KEMPE: *Handbook of Electrical Testing*, p. 166; SILVANUS THOMPSON: *Elementary Lessons on Electricity and Magnetism*, p. 318.

[4] An astatic system of needles consists of two needles suspended parallel to each other and near together, with their poles placed in contrary directions. One is a little

Galvanometer for the Wheatstone Bridge. 199

with a permanent magnet, called a *directing magnet*, by which the needle can be brought to zero in any desired angular position of the apparatus.

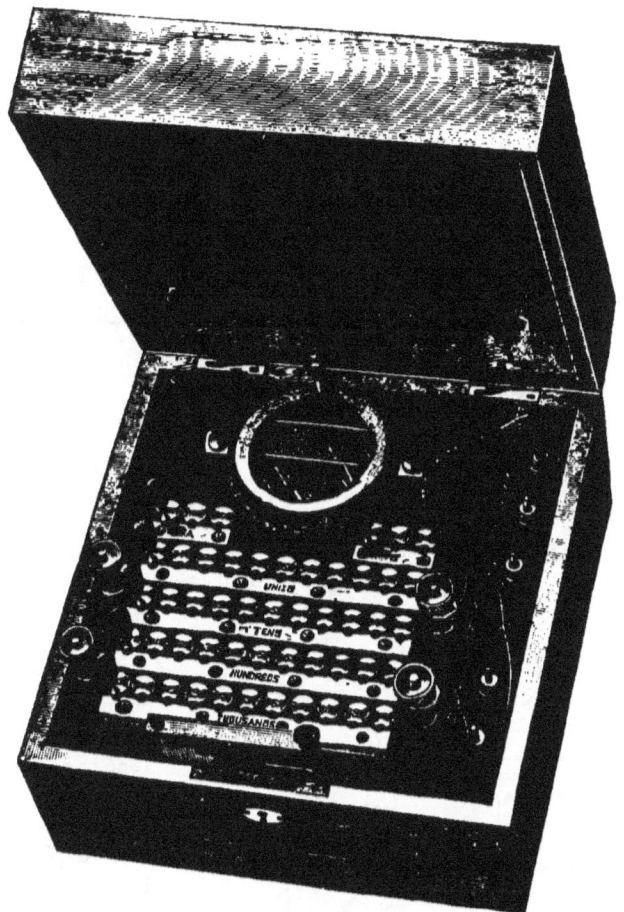

FIG. 172. Wheatstone Bridge Apparatus, with Galvanometer and Battery.

346. To Measure the Conductivity Resistance of a Telegraph Line.—Have the remote end of the line put to ground, taking care that no relays are left in circuit. Connect the home end

stronger than the other, so that the pair has a very feeble tendency to place itself in the magnetic meridian (86 *b*). Such a system is capable of being deflected by a very weak current, and hence is used in the construction of the more delicate types of galvanometers.

200   *Testing Telegraphic Lines.*

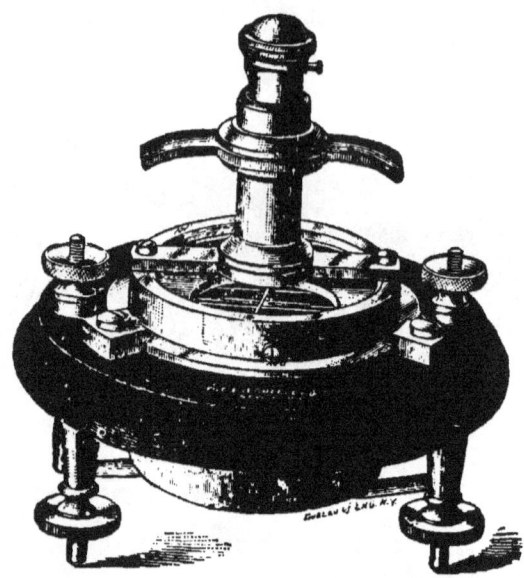

FIG. 173. Astatic Galvanometer with Directing Magnet.

of the line to the terminal C, and the terminal E to the ground, as in Fig. 174. Unplug from A B (*b*) and B C (*a*) each, a resistance most nearly approximating in value that of the line to be measured (342). Usually this will be 1000 in each. Press the right-hand or battery-key B' and remove plugs from E A (*d*) until the resistance unplugged equals roughly that which is to be measured. Then depress the left-hand or galvanometer-key A', and rearrange and adjust the plugs in E A (*d*) until the key can be repeatedly opened and closed without causing any movement of the

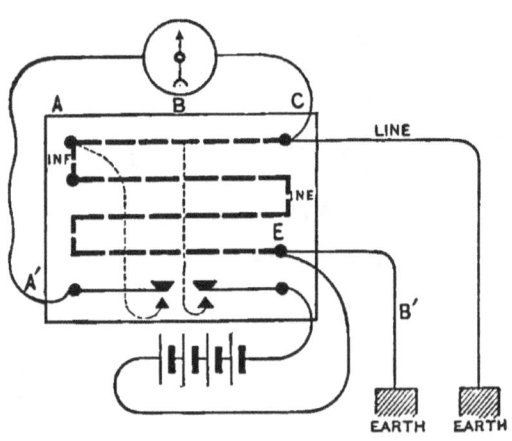

FIG. 174. Conductivity Test.

galvanometer-needle.[5] When this balance has been effected, the resistance unplugged in E A is equal to the conductivity resistance of the line under test.

346a. **Conductivity Resistance by Loop Method.**—A more accurate method of making conductivity tests, which is available whenever there are three or more parallel wires available between the same points, is the *loop test*. If we suppose the wires to be Nos. 1, 2, and 3, they are looped together in different combinations at the distant station, and the resistances of the several loops taken in succession, one end of the loop under test being connected to C and the other to E, as in Fig. 171. For example, suppose the resistances to measure as follows:

```
Resistance of 1 and 2, looped............  6550 ohms.
     "       2 and 3,   "    ............  6180  "
     "       1 and 3,   "    ............  6830  "
                                           ─────
     Sum of all three........................ 19560  "
```

As by this process the resistance of each wire has been taken twice over, we divide the amount by $2 = 9780$ ohms.

If we deduct from this result the total of each pair of looped wires in succession, the remainder in each case must be the resistance of the wire not in the loop. Thus:

$9780 - 6180 = 3600$ for No. 1 wire.
$9780 - 6830 = 6830$ for No. 2 wire.
$9780 - 6550 = 3230$ for No. 3 wire.

Conductivity tests may also be made with sufficient accuracy for most purposes, and with great convenience and facility, by means of a properly constructed voltmeter. See (369).

347. **Earth Currents.**—In making conductivity tests with the distant end of the line to ground (346), interference is sometimes caused by *earth currents*, which flow through the wire, and aid or oppose the testing current, as the case may be. When these are tolerably steady and not too strong, their effect may be eliminated by making measurements with both a positive and a negative current and taking the mean of the two results. Sometimes, however, these earth currents are so strong that an accurate measurement cannot be made, and the loop method (346a) must be resorted to.

---

[5] Care should be taken in all cases, when finally closing the circuit by the second or galvanometer-key, to first make very short contacts or " taps," just enough to indicate the direction of the deflection of the needle, until the coils are nearly adjusted to a balance, otherwise much time will be needlessly lost by the oscillations of the needle.

348. Measurement of Resistance of Ground Plate at Distant Station.
—The principle of this measurement is the same as that last described. Two lines are necessary; the earth takes the place of the third line. For example, suppose we have:

Resistance of No. 1 and No. 2, looped......	6550 ohms.	
"	" No. 1 and distant ground...... 3625	"
"	" No. 2 and distant ground...... 2975	"

Half the sum of which is 6575, from which deduct loop resistance of Nos. 1 and 2, gives 25 ohms as resistance of ground. The resistance of a ground plate ought not to exceed 10 ohms.

349. Measurement of Insulation Resistance of a Line.
—In making this test, the connections at the home station are the same as in the conductivity test (346); the line is *open* or insulated at the distant station, instead of being to ground. In most cases, the insulation resistance will exceed the amount of resistance available in the E A side of the bridge. In this case the resistance in · A B (*b*) must be made greater than that in B C (*a*). For example, it may be 10 in B C and 100 or 1000 in A B. In this case, when the balance has been obtained, the amount unplugged in E A (*d*) must be multiplied by 10 or by 100, as the case may be, in order to obtain the correct resistance. It will be observed that under these conditions, the *ratio* of the resistances in the different parts of the bridge remains unchanged.

350. Location of the Position of a Ground.
—When the fault is a *dead ground*, which is not often the case, it is a very simple matter to locate it. For example, if the line were 250 miles long, and from previously recorded measurements its conductivity was known to be 3250 ohms, or 13 ohms per mile, and the resistance measured through the fault was 1287 ohms, then the distance from the testing station would be $1287 \div 13 = 99$ miles.

351. Location of the Position of an Escape.
—This is one of the most common cases which arise in practice. If no other than the defective line is available for the measurement, the process presents some difficulties, for the reason that the resistance of the fault is usually variable. If we have, for example:

(1) Resistance of line in good order (from previous tests).. 4500 ohms.
(2)   "   with distant end open (measured)............ 3500   "
(3)   "   with distant end to ground (measured)...... 2700   "

Subtract (3) from (2) and also (3) from (1); multiply the two remainders together and extract the square root of the product, and finally subtract this result from (3). In the above case, this would give the resistance of the conductor to the fault as 1500 ohms. While this method is theoretically accurate, it will not do to depend too much upon it in practice, for the reasons given.

352. **Method of Double Measurement.**—Let two measurements be made, one from each end, the opposite end of the line being open. Suppose the fault the same as in the last case. By records and measurements we have,—

(1) Conductivity resistance of wire when good.... 4500 ohms.
(2) Resistance measured from A with B open..... 3500 "
(3)   "   "   "   B with A open..... 5000 "

To find the resistance from A to the fault.—Subtract (3) from the sum of (1) and (2), and divide the remainder by 2.
To find the resistance from B to the fault.—Subtract (2) from the sum of (1) and (3), and divide the remainder by 2.

The fault is 1500 ohms from A and 3000 ohms from B, which may be reduced to miles as in (350).

353. **The Loop Test.**—When a good wire is available between the same points as the defective wire, this method may be made to give extremely accurate results in the hands of a careful operator. The arrangement of the connections, the method of making the measurements, and the computation of the result are precisely the same as in the method described for measuring a distant ground in (346). If the resistance of the fault is considerable, care should be taken to employ sufficient battery-power to get decided deflections on the galvanometer. The loop should be made at the nearest station available beyond the fault.

354. **Varley's Loop Test.**—The arrangement of connections for this modification of the loop test is shown in Fig. 175. The defective wire is looped with a good wire, and terminal B is connected to a grounded battery. B C and A B are the fixed resistances; E A is adjusted until equilibrium is reached. The actual connections are shown in Fig. 176. The calculation is made as follows:

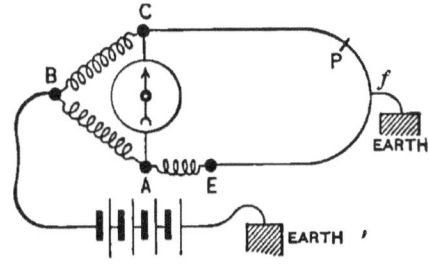

FIG. 175. Principle of Varley's Loop Test.

Suppose a line having a total conductivity resistance of 4500 ohms looped with another line of the same resistance, we should then have:

(1) Total resistance of loop.................. 9000 ohms.
(2) Resistance in E A to balance............. 6000 "

Subtract (2) from (1) and divide remainder by 2, gives number of ohms between terminal E and the fault (in this case 1500), which is reduced to miles in the usual way (350).

The defective wire must always be attached to terminal E, or the needle cannot be made to balance. Therefore, in case a balance cannot be obtained, the obvious remedy is to reverse or interchange the loop connections with the rheostat.

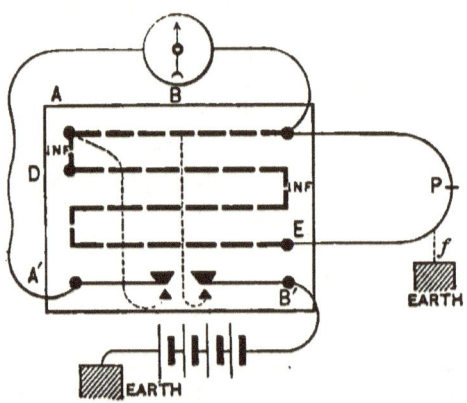

FIG. 176. Varley's Loop Test.

355. To Locate a Cross. — In case a third good wire is not available, connect one of the crossed wires to C and the other to E of the bridge. Make one measurement of the loop through the cross with both lines open at nearest available station beyond, and another with the same wires looped at that station. If the two measurements are approximately the same, the number of ohms in the loop divided by 2 and converted into miles will give the distance of the cross from the testing station. If the lines are of different length, owing to the routes being different, allowance must be made for the fact; also, if the wires are of different conductivities per mile.

356. When the two measurements differ considerably, showing that the cross offers more or less *resistance*, the above test would give a result in excess of the real distance. In such case the following procedure may be adopted. In Fig. 177, suppose wires No. 1 and No. 2 to be crossed at X.

By measurement from A B we get the following results, for example:

(1) No. 1 from A to C (with No. 2 open at B and D)....... 3000 ohms.
(2) No. 2 and No. 1 from A to B through the cross at X... 4650 "
(3) No. 2 and No. 1 from B to C through the cross at X... 2650 "

Shunts of Galvanometers. 205

Deducting (3) from the sum of (1) and (2) gives 5000, which divided by 2 gives 2500 ohms, as resistance of No. 1 wire from A to X. The distance on No. 2 wire may be found, if desired, in the same way.

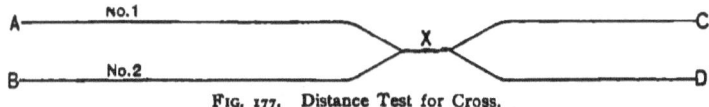

FIG. 177. Distance Test for Cross.

357. When a third wire in good order is available, the most convenient as well as the most accurate method of locating a cross, is to ground either one or both ends of one of the crossed wires and make the other crossed wire into a loop with the good wire. The cross can then be treated as a *ground*, and located by one of the loop tests heretofore given in (353) and (354).

358. **To Locate a Bad Joint or Abnormal Resistance.** —It sometimes happens that a line gives a much higher resistance than it should do, according to computation or by previous measurement. In such cases a *bad joint* may be suspected. To locate it, instruct a station midway of the line to put on ground. Take a measurement through first half of the line and this ground, the distant end being open. This will show whether the fault is in the section measured or beyond. Repeat the test to another station in the middle of the defective section, and so on until it has been fixed between two sections.

359. **Measurement of very High Resistances.**—The highest resistance which can be measured by the Wheatstone Bridge apparatus, described in (344), is 1 megohm, or 1,000,000 ohms. This is a sufficient range to cover most of the requirements ordinarily met with in practical telegraphy, but in testing insulators, or the insulation resistance of very short sections of out-of-door line, it is often desirable to be able to determine much higher resistances. The method of *proportional deflections* is usually resorted to in such cases. A galvanometer having a coil of a large number of turns of very thin wire and a delicately suspended needle (345) is most suitable for the purpose.

360. **Shunts of Galvanometers.**—The galvanometers for this work must be provided with *shunts;* these are short coils of wire, arranged to be connected or bridged across the terminals of the galvanometer, and are usually marked (to indicate their multiplying power) 1·10, 1·100, and in very delicate instruments usually 1·1000 also.

The first shunt coil has 1-9, the second 1-99, and the third 1-999 of the resistance of the galvanometer coil. They are made of copper wire, that they may be affected by temperature in the same ratio as the galvanometer coils. Fig. 178 shows an arrangement much used, in which either of three shunts may be thrown into use at will by changing the peg.

FIG. 178. Shunt Box for Galvanometer.

361. Measurement by Deflections.—This method is useful in making comparative tests of insulators. In this case the internal resistance of the testing battery is inappreciable in comparison with the resistance to be measured, and hence the force of the current acting upon the needle may, without sensible error, be regarded as proportionate to the $e.\,m.f.$ of the battery.

First, connect the galvanometer G in circuit with a large known resistance R (say 10,000 ohms) and a single cell E, whose $e.\,m.f.$ is known, as, for instance, a gravity cell (9). If the deflection exceeds 12°, reduce it to a point below that figure by the use of the proper shunts.[6] The arrangement of the connections for performing this operation, which is termed *taking the constant* of the galvanometer, is illustrated in Fig. 179. *Second*, remove the shunt and the resistance R, and having replaced the latter by the unknown resistance to be measured, add a sufficient number of cells of the same kind (in series) to produce a convenient deflection, not exceeding 12°, as before. The result is found by simple rule of three, as in the example given in the next paragraph.

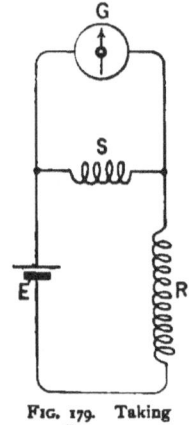

FIG. 179. Taking Constant.

362. Measurement of Resistance of Insulators.—Mount a set of say 10 insulators, I, Fig. 180, upon a suitable frame out of doors, exposed to rain under the same conditions as if in actual service. Bind a line-wire to the whole series, and connect this with one terminal of the galvanometer G, the other terminal of

---

* The reason for this procedure is, that above this point the angles of the deflections cease to be proportional to the strength of the currents producing them. (Compare table of tangents, p. 55.)

galvanometer to zinc pole of battery E, and the copper pole of battery to ground. Suppose that with the particular galvanometer used, the following results are obtained, the weather being very wet:

    1 cell through 10,000 ohms............................ 41°
    1 " " 10,000 " (with 10 shunt).............. 5°
    Constant of galvanometer (1 cell through 10,000 ohms).. 50°
    10 cells through 10 insulators in parallel................ 10°

Therefore, if 1 cell will give 50° through 10,000 ohms, as per constant, 10 cells will give 50° through 100,000 ohms; and will therefore give 10° through 500,000 ohms.

Hence the joint resistance of the 10 insulators is 500,000 ohms, and their mean individual resistance 5,000,000 ohms, or 5 megohms, per insulator.

363. Measurement of the Internal Resistance of a Battery.—(1) It follows from Ohm's law (124) that when the total resistance of any circuit, embracing that of an included battery and galvanometer, is *doubled*, the quantity of the current flowing through it is *halved;* and hence if the indications of the galvanometer be proportional to the strength of current, its deflection will also be halved. If a tangent galvanometer (96) of known resistance is at hand, connect it with a plugged rheostat in the circuit of the battery to be measured. Reduce the sensitiveness of the instrument by a shunt (360), if necessary, to bring the deflection as near 60° as possible, and note the corresponding tangent of the deflection as given in the table, p. 55. Unplug resistance until the tangent of the deflection is halved, showing that the total resistance has been doubled. Deduct the resistance of galvanometer and connections from the added resistance; the remainder is the resistance of the battery. Do not forget that the shunt, when used, diminishes the resistance of the galvanometer as a part of the measured circuit.

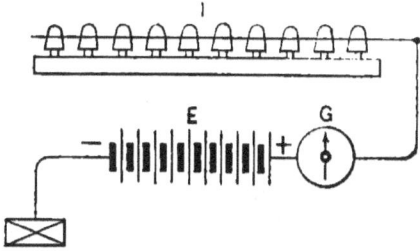

FIG. 180. Test of Insulators.

364. (2) If the resistance of the galvanometer is unknown, a modification of the Wheatstone bridge may be used. Make the connections as in Fig. 181. Connect the terminals B' and E by a short, thick wire. The left (galvanometer) key is permanently depressed. Touch the right-hand key and adjust resistance A E (*d*) until the needle remains at rest (it will not be at zero). It is neces-

sary to shunt the galvanometer in this test. If the resistance in A B (*a*) is equal to that in A B (*b*), the amount unplugged in A E is equal to the resistance of the battery. In any case, by proportion, A B is to A C as A E is to the resistance of the battery.

The most accurate results will be reached when A B is as high and A C is as low as possible, but not so high as to carry A E beyond the range of the rheostat.

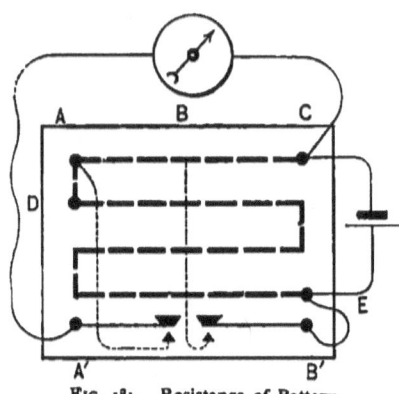

FIG. 181. Resistance of Battery.

365. **Measurement of Resistance of Galvanometer.**—If, in the diagram, Fig. 181, the battery and galvanometer are made to change places, the resistance of the galvanometer may be determined in the same way. Make B C (*a*) not more than one-tenth of the probable resistance of galvanometer, and make A B (*b*) not less than ten times the same, but not so high as to carry A E beyond the range of the rheostat. The least possible value of B C with an ordinary bridge set would be 10 ohms. A smaller resistance might be extemporized from a piece of wire, if necessary.

366. **The Differential Galvanometer.**—This instrument is primarily designed to show the difference in strength between two currents. The coil is wound throughout with two wires, equal in length, resistance, and number of convolutions, so that the same current in each will have a like effect upon the needle. The two wires are sometimes formed into a tape by plaiting together the silk with which they are covered. If, therefore, two equal currents traverse the respective wires in opposite directions, the needle will not move. If one current be stronger than the other, the needle will be moved by the stronger current with a force due to the difference in the strength of the two currents. This instrument was formerly much used to measure resistances by comparing them with standard resistance coils, but has now been practically superseded by the Wheatstone bridge (341).

367. **Testing for Insulation by Received Currents.**—This system of testing offers many advantages over that hereinbefore referred to (340) for the daily examination of telegraph lines. The current from a testing battery, of a definite and approximately uni-

Testing for Insulation by Received Currents.

form $e.\ m.\ f.$, is sent at a stated time through the different lines, or sections of lines, and the volume of current, as indicated by a tangent galvanometer such as that shown in Fig. 182, or by an ammeter (369) at the receiving end, is registered. It is evident that the strength of the received current will be greater or less as the insulation is better or worse, and hence if the $e.\ m.\ f.$ of the battery be constant, the volume of the received currents as observed from day

FIG. 182. Western Union Tangent Galvanometer.

to day will give an accurate knowledge of the condition of the lines. The normal resistance of each line is known from the stated conductivity tests, and so if the currents be sent from a battery of known $e.\ m.\ f.$ it is only necessary to divide the latter amount by the former to know at once the maximum current which can possibly be received through any wire. For example, if a battery of 50 cells be used on a circuit of 2500 ohms, then

$$50 \text{ volts} \div 2500 \text{ ohms} = 0.02 \text{ ampère}.$$

This may be regarded as the standard current of that circuit, and the greater the leakage the greater will be the diminution of the current below that standard. Tables may be made for convenient reference showing the normal current of each line. Special faults are of course investigated by the bridge apparatus (341) when their presence has been revealed by the procedure above described.

368. Use of the Voltmeter and Ammeter in Telegraphic Testing.—Since the general introduction of electricity in lighting and power service, a new class of instruments for the measurement respectively of potential and current, known as *voltmeters* and *ammeters*, have been brought to great perfection, and are now frequently employed with advantage in telegraphic work. Much time is saved in making readings and computations, as a simple inspection of the indication of the pointer on the scale at once gives the result in volts or ampères. The pointer comes to rest promptly, so that a reading can be made almost instantaneously.

369. The Weston Ammeter and Voltmeter.—Fig. 183 shows Weston's portable type of direct-reading mil-ammeter, about one-fourth its actual size.

FIG. 183. Weston's Mil-ammeter.

It comprises a permanent magnet, having a hollow rectangular coil of aluminium wire suspended within its field upon jeweled pivots, to which coil the pointer is attached. Fig. 184 is a full-sized view of the working parts of the instrument. One of the most useful types for telegraphic work has a scale reading to 1 ampère, with subdivisions of 10 mil-ampères, which may be read by inspection to a single mil-ampère.[7] Another type, which is adapted to perform all the measurements ordinarily required in a large telegraph station, is provided with several scales; one of a single volt, which may be read to .001 volt, for determining the potential of a single cell; another of 0 to 500 volts (which can be read to single volts or half volts), for taking the potential of a large number of cells when connected in series in a single battery; another of 1 ampère (which can be read to mil-ampères) for determining the strength of currents. Some are made

---

[7] The construction of the voltmeter and ammeter are similar, the difference being in the length and thickness of the wire in the deflecting coil, which is made long, thin, and of great resistance in the voltmeter, and comparatively short and thick and of small resistance in the ammeter. Two or more coils of different lengths may be fitted to the same instrument, as in a galvanometer, giving different grades of sensibility.

The Weston Ammeter and Voltmeter.

with a coil of precisely 100 ohms resistance, giving a full scale deflection with 1 volt. Such an instrument is very convenient for measuring line resistances. For example, with 100 volts the resistance in circuit required to bring the pointer to the upper division of the scale would be 10,000 ohms, and hence by pointing off two decimal places any resistance in ohms in the circuit may be determined by direct reading from the scale, in the same manner as volts.

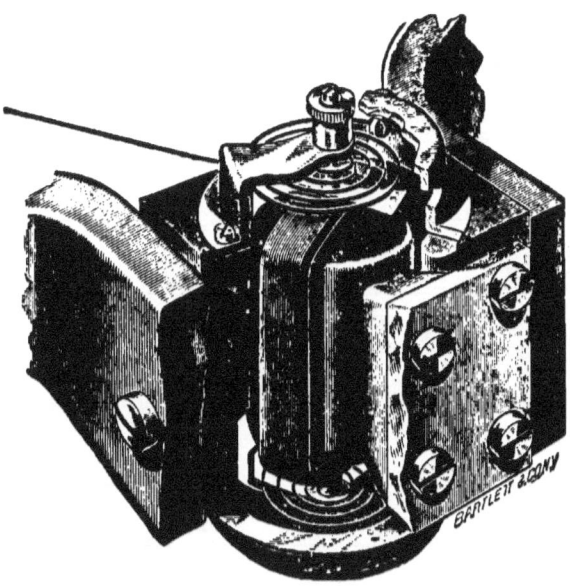

FIG. 184. Mechanism of Weston's Direct-Reading Instrument.

The Weston instruments are particularly well adapted for all current measurements usually performed with a tangent galvanometer (102). They are not only, for most purposes, more accurate, but are far more convenient, as they may be placed in any position, and are in no wise affected by the neighborhood of masses of iron or of foreign electric currents. No time need be lost in leveling, adjusting, or waiting for the needle to settle, while the convenience of being able to read off the results directly without calculation is very great.

Another and a very important advantage is, that the tests may be made with the same current which is employed in the ordinary operation of the circuit. Tests for resistance, especially, not unfrequently give very fallacious results, when made, as is often the case,

Western Union Telegraph Company.

Test of Conductivity at _____ 18___, Weather _____ Therm. _____ Deflection of Galvanometer through _____ Ohms

Number and Terminus.	Station Tested with.	Miles Copper Wire.				Miles Plain Wire.				Miles Galvanized Wire.				Total Miles.	Relays in Ckt.		Deflec. of Galv'r through Wire.	Resistance in Ohms.		Remarks.
		No. 11.	No. 11½.	No. 12.	No. 14.	No. 6.	No. 6.	No. 8.	No. 10.	No. 6.	No. 6.	No. 8.	No. 10.		Number.	Resistance.		Total.*	Mileage.*	

* These columns will be filled in at the Electrician's Office, when measurements are made with Tangent Galvanometer.
(NOTE.)—The Tests for Conductivity should be made in dry weather, one day in each week.

Western Union Telegraph Company.

(No. 83.)

Test of Insulation at _____ 18__, Weather _____ Deflection of Galvanometer through _____ Ohms _____

Number and Terminus.	Station Tested with.	Thermometer.		Miles of Insulation.			Deflec. of Galv'r through Wire.	Resistance in Ohms.		Remarks.
		Dry Bulb.	Wet Bulb.	Glass and Pin.	Kenoba.	Sq. Glass.	Total Miles.	Total.*	Mileage.*	

* These columns will be filled in at Electrician's Office, when measurements are made with Tangent Galvanometer.

with a current of much lower potential than the actual working current.

370. **Recording Tests of Conductivity and Insulation.**—The forms of returns for line tests adopted by the Western Union Telegraph Company are given on pp. 212, 213. When the tangent instrument is used, the constant (361) is written in the upper right-hand corner of the sheet. One horizontal line is appropriated to each separate wire tested. The headings sufficiently explain entries to be made in the several columns. When the tests are made by the bridge apparatus, the results are entered directly in the resistance columns, but if with the tangent instrument, these are computed and filled up at the electrician's office in New York. The same observations apply to the insulation form. The record of the wet and dry bulb thermometer is important, as it enables the percentage of moisture in the air to be determined and its effect upon the different kinds of insulation to be compared and studied.

By inspecting and comparing these sheets, as returned from the various testing offices, the electrician's department is kept fully informed of the electrical condition of the lines in all parts of the country. The system of stated reports was instituted by Lefferts, of the American Telegraph Company, in 1863,[8] and has resulted in a vast improvement in the efficiency of the service.

---

[8] LEFFERTS (MARSHALL), born January 15, 1821, in Bedford, now part of Brooklyn, N. Y. In early life he was a civil engineer and was employed in laying out the city of Brooklyn. Subsequently he became a successful merchant in New York City, and a prominent militia officer. In 1849 his marked scientific tastes led him to become interested in telegraphy. Entering into the new enterprise with the energy and zeal which were among his most notable characteristics, he organized and became the president and manager of a range of lines operating the Bain electro-chemical system, extending from New York to Boston and Buffalo. Legal complications in connection with patents eventually led to a consolidation of these lines with those controlled by the Morse patentees, in consequence of which he resumed for a time his manufacturing and mercantile business. In 1860 he was appointed engineer and executive manager of the American Telegraph Company, which under his administration became one of the most popular and successful telegraph organizations that ever existed on this continent. He retained this position until the consolidation of the American with the Western Union Telegraph Company in 1866, and subsequently occupied a responsible post in the united service until 1871. At this date he was elected president of the Gold and Stock Telegraph Company of New York, which position he held until his death. He possessed a most unusual organizing and executive ability, and while a strict disciplinarian, it afforded him genuine pleasure to discover and to reward meritorious service, even in the humblest capacity. His uniformly just and considerate treatment of his employees, no less than his genial and kindly spirit, insured the most loyal, enthusiastic, and diligent service from all. By a liberal system of advancement to the intelligent, the skillful, and the deserving, the standard of character and acquirements among the employees of the American Company was elevated to an extent to which later times have afforded few parallels. As engineer of

this extensive organization, he labored unceasingly to place its service upon a permanent foundation befitting its importance and its high mission. He was the first to appreciate the importance of testing lines and apparatus, and it is to the standard of excellence which he established that the commencement of the era of scientific telegraphy in America may be traced. Assuming in 1861 the administrative management of a heterogeneous assemblage of poorly built and ill-arranged telegraphs, equipped with a miscellaneous collection of apparatus of antiquated and unserviceable types, he five years later turned over to the Western Union Company 30,000 miles of wire, constituting perhaps the most complete, thoroughly organized, and efficient telegraphic system in the world. The influence of the reforms and improvements which were instituted during his administration will continue to be felt in the American telegraph service for all future time. He died suddenly, July 3, 1876, while on his way, as commander of a military organization, to participate in the celebration of the centennial anniversary of the Declaration of American Independence, in Philadelphia.

CHAPTER X.

HINTS TO LEARNERS.

371. Formation of the Telegraphic Code.—The code of alphabetical and numerical signals employed in telegraphy, as devised by Vail in 1837,[1] is made up of various combinations of a small number of elements. In the so-called "Morse" code, as used in America, there are seven of these elements, viz.:

(1) The *dot;* (2) the *dash;* (3) the *long dash;* (4) the *ordinary space;* (5) the *letter-space;* (6) the *word-space;* and (7) the *sentence-space.* It is important to remember that the value of the spaces in the code is as great as that of the dots and dashes. A common misconception exists in the minds of students that the telegraphic code consists exclusively of dots and dashes. The foundation of perfect telegraphic manipulation lies in the ability, which can only be acquired by careful observation and training, to accurately divide and subdivide *time* into intervals which are multiples of an arbitrary unit.

---

[1] VAIL (ALFRED), born at Speedwell, near Morristown, N. J., September 25, 1807. In early life he became an apprentice in his father's Speedwell iron-works. After attaining his majority, he pursued a course of study and graduated at the University of the City of New York with the intention of entering the ministry, but in September, 1837, chancing to witness one of the early experiments of Morse with his crude telegraphic apparatus, his mind, naturally of a strongly scientific cast, was instantly fired with enthusiasm at the future possibilities of this marvelous invention. He became wholly absorbed in the enterprise, and persuaded his father, Stephen Vail, to furnish the means required to perfect, develop, and introduce the electric telegraph. Among the improvements in the apparatus and methods originated by himself, of the utmost practical value, were the register (269), which is to-day but little changed from the form he gave it in 1844, and the "Morse" alphabetical code (372), now in universal use in America. (See American Inventors of the Telegraph, *Century Magazine,* xxxv. 924, April, 1888.) The efforts of Vail in overcoming the numerous practical difficulties that beset the work of installing the pioneer telegraph line between Washington and Baltimore in 1844, were indefatigable, and it is to his genius, patience, and untiring diligence that the ultimate success of the enterprise was in no small measure due. The last ten years of his life were passed by him in comparative retirement, engaged in his favorite pursuits of science and literature. He died at Morristown, January 18, 1859.

The American Morse Code.

372. The American Morse Code.—The complete code as now used in the United States and Canada, comprising letters, numerals, punctuation, and other signs more or less used, is given below:

I. ALPHABET AND NUMERALS.

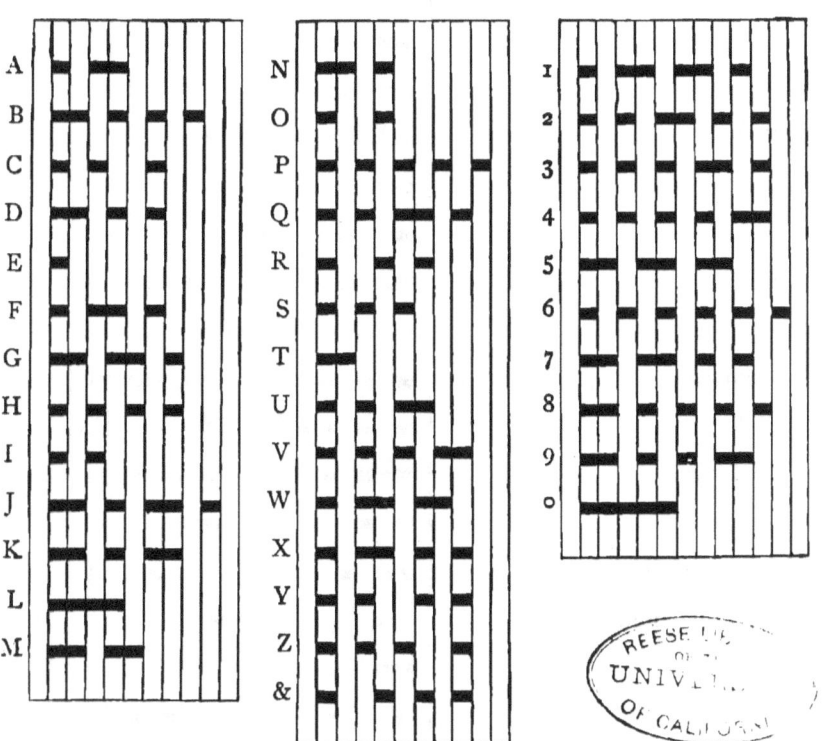

The arbitrary unit of time in this code, which, when written down, becomes a unit of length, is technically termed the *dot;* an unfortunate name for this element, inasmuch as it conveys the idea of an inappreciable lapse of time, or of the transmission of a current of infinitely short duration. On the contrary, an appreciable *time* is required for the production of signals by electricity (315); in the magnetization of electro-magnets (195), and in the movement of clock-work. The formation of a dot, therefore, necessarily involves *time*. Assuming, therefore, that

(1) The *dot* is the unit of time,
(2) The *dash* is equal to 2 dots;

II. PUNCTUATION, Etc.

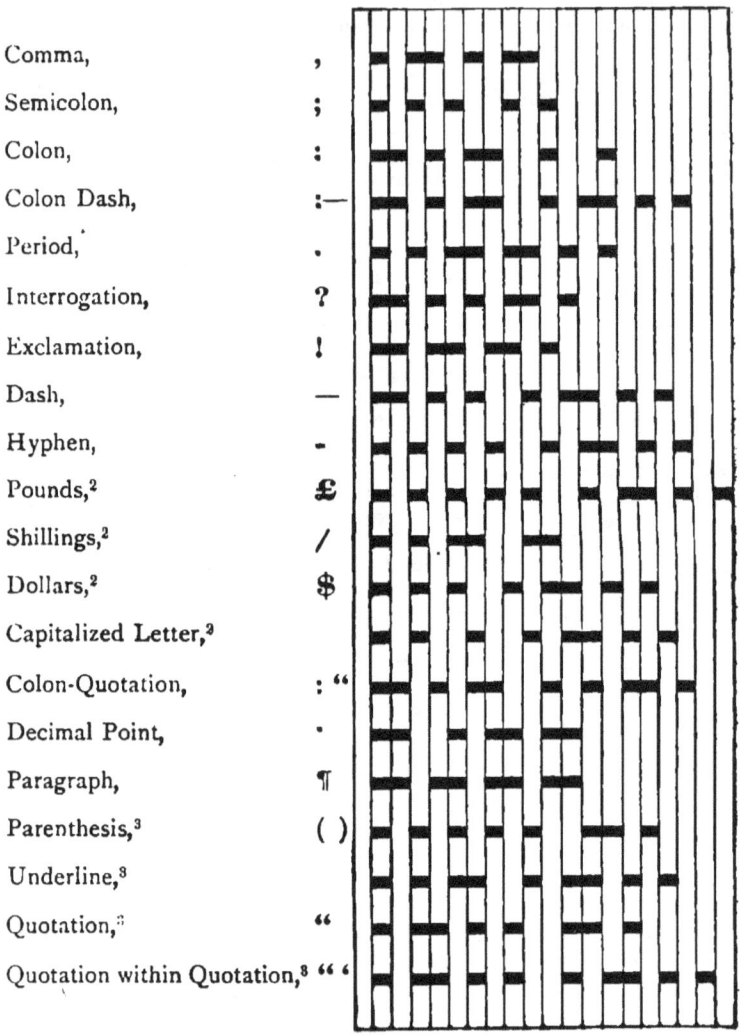

Comma,	,
Semicolon,	;
Colon,	:
Colon Dash,	:—
Period,	.
Interrogation,	?
Exclamation,	!
Dash,	—
Hyphen,	-
Pounds,[2]	£
Shillings,[2]	/
Dollars,[2]	$
Capitalized Letter,[3]	
Colon-Quotation,	: "
Decimal Point,	.
Paragraph,	¶
Parenthesis,[3]	( )
Underline,[3]	
Quotation,[3]	"
Quotation within Quotation,[3]	" '

(3) The *long dash* is equal to 4 dots;
(4) The *ordinary space* between the elements of a letter is equal to 1 dot;
(5) The *letter-space* is equal to 2 dots;
(6) The *word-space* is equal to 3 dots;
(7) The *sentence-space* is equal to 6 dots.

[2] To be used before the characters to which it refers.
[3] To be used before and after the words to which it refers.

The old rule in transmission was to make the dash equal to 3, and the long dash to 6, dots. When the receiving was largely done by recording instruments this was a most necessary requirement, for a dot, and a dash equal to only 2 dots, might easily be mistaken for each other in reading by sight, but now that receiving by sound has become practically universal, this objection has lost its force, and by shortening the dashes a material gain in rapidity of transmission is effected without any corresponding disadvantage.

373. **Learning the Code.**—The student should first thoroughly commit to memory the groups of signs representing the letters of the alphabet, the numerals, and the principal punctuation points, viz., the *period*, the *comma*, and the *point of interrogation*. The remaining characters can be learned afterwards, as they will be little needed by the beginner.

374. **Handling the Key.**—The most approved manner of grasping the key, and one which has been employed by some of the most successful, experienced and rapid American operators, is shown

FIG. 185.

in Fig. 185. Curve the fore-finger, but do not hold it rigid. Let the thumb press slightly in an upward direction against the knob. Keep the wrist well above the table. No better general direction can be given than that the key should be grasped, held, and controlled with the same flexible but perfectly controlled muscular action of the fingers, wrist, and fore-arm with which the skilled penman holds his pen. Carefully avoid *tapping* upon the knob of the key; the raising spring should assist the upward motion of the key, but should never be permitted to control it.

By constant drill, as hereinafter directed, the habit of making dots with regularity, uniformity, and precision must first be acquired; then dashes, and lastly in order, group of dots and dashes, letters and words. If possible for the student to obtain a register (269), he should by all means employ it in his practice, for he will then be more easily enabled to observe and correct the faults in his own manipulation. In commencing, the habit should at once be acquired of making the dots like short, firm *dashes*. The student should learn to form the conventional characters accurately and perfectly; speed will come in good time, but only as the result of constant and persistent practice, accompanied by a determination to excel.

375. **Elementary Principles of the Code.**—As a basis for practice, the code may be regarded as comprising six elementary principles, viz.:

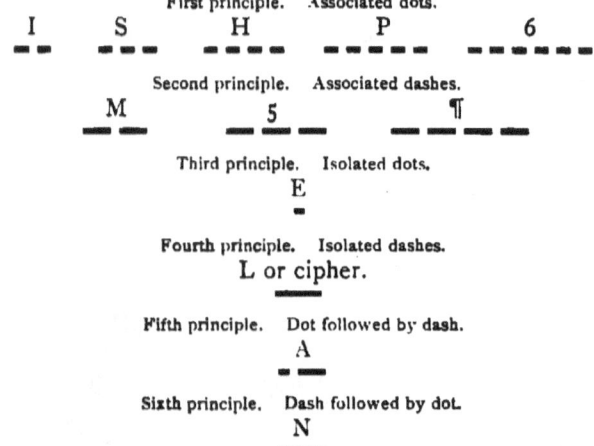

376. **Preliminary Practice with the Key.**—The student should first practice upon the above elementary principles.

(1) Make dots with the key at uniform and regular intervals, until they can be produced with the precision of a machine, and of definite and uniform dimensions. The student will find this more easy, if at first he times himself by the beats of a watch or a small clock.

(2) Next, make *dashes*, first at the rate of about *one* per second, which speed may be increased by degrees, as skill is acquired by practice, to *three* per second. Make the space interval between successive dashes as short as possible. If the upward movement which forms the space be made full, it cannot be made too quickly.

(3) The third principle occurs but once, and needs no specific directions.

(4) This principle will be found somewhat more difficult to execute. The usual tendency is to make T too long, and L too short. Theoretically, the cipher is one-half longer than L, but in fact it is always made the same, as the practice has been found to occasion

Preliminary Practice with the Key. 221

no inconvenience. Occurring alone, or among other letters, it is translated as L, but when found among figures is read as o.

(5) The fifth principle forms the letter A. The usual tendency is to separate the two elements too much.

(6) The dash followed by a dot (N) is usually found to be somewhat difficult. Time the movement by pronouncing the word *ninety*, sounding the first syllable fully. Guard especially against the usual tendency to separate the elements by too great a space.

377. Exercises upon Code Characters.—Having become thoroughly familiar with the principles, the following exercises may with advantage be taken up in order:

(1) E   I   S   H   P   6

These should be practiced repeatedly until the correct number of dots in each character can be *certainly* made at every trial. A habit once formed of making the wrong number, usually one or two too many in the case of H, P, and 6, is almost impossible to eradicate. Guard especially against the objectionable habit of shortening or *clipping* the final dot, a vice which leads to innumerable and vexatious errors and misreading of signals.

(2) T   M   5   ¶

The faults to be particularly guarded against in this exercise are shortening or elongating the terminal dash, and separating the successive dashes by too great a space interval.

(3) A   U   V   4

The usual tendency to allow too much space between the dot and dash in the above letters may be overcome by forming them as by an elongation of the final dot in I, S, H, and P.

(4) I   A   S   U   H   V

Practice these characters in pairs, that the distinction between them may be more firmly impressed upon the mind.

(5) N   D   B   8

The student who has mastered the sixth principle will find no difficulty with the above characters.

(6) A   F   X   ,   W   1

(7) U   Q   2   Period   3

These are similar to preceding exercises and present no new difficulties.

(8) K   J   9   ?

     G   7   Exc.

J and K are usually considered the most difficult letters in the code. Avoid the tendency to separate J by a space into double N, and be careful that the dashes are of equal length. The numerals 7 and 9 require some care to ensure correct spacing.

(9) O   R   &   C   Z   Y

These are termed the *spaced letters*, and the utmost care and diligent practice are necessary in order to form them accurately. *The ability to transmit the spaced letters with absolute correctness is the test of a strictly first-class sender.* The space should be just enough in excess of that ordinarily used between the elements of a letter to enable the letters intended to be made to be distinguished with certainty from I, S, and H. The most usual tendency is to make the space too great, even in some cases as great as the space between letters. This is a most fruitful source of misapprehension and error, and too much pains cannot be taken to acquire and maintain correct habits in this particular.

In transmitting words containing groups of two or more spaced letters, careful operators are accustomed to slightly increase the spacing *between the successive letters of the group.*

Practice in transmission from miscellaneous manuscript is strongly recommended. The ability to read all kinds of copy; good, bad, and indifferent, correctly at sight, is a most valuable one, and is not difficult to acquire by attention and experience.

If the principles here laid down be firmly adhered to, the learner will find much reason for encouragement, not only at the rapidity with which he will master what at first sight appears to be a very difficult undertaking, but at the extreme accuracy with which he will be able to manipulate his instrument after a fair amount of practice. He must also carefully bear in mind that one of the most universal faults among those attempting to learn the telegraphic art, is that of going over a great deal of ground and learning nothing thoroughly.

378. **Reading by Sound.**—This art can only be acquired by constant and persevering practice, keeping in mind the principles above given. The lever of the telegraphic sounder makes a sound at each movement, the downward motion producing the heavier one. The *down-stroke* indicates the commencement of a dot or a dash and the *up-stroke* its termination. A dot makes as much sound as a dash; the only difference is in the length of time or *interval* which elapses between the two successive sounds. Thus, if the recoil or up-stroke were absent, it would be impossible to distinguish E, T, and L from each other.

379. In learning to read by sound it is advisable for two persons to practice together, taking turns at reading and writing, and each correcting the faults of the other. The sounds of the code characters must first be learned separately, and then short words chosen, which must be written very slowly and distinctly and well spaced, the speed of manipulation being gradually increased as the student becomes more proficient in reading. After becoming sufficiently well versed in the art to read at the rate of twenty-five or thirty words per minute, further practice may best be had in copying with a *pen and ink* (not with a pencil) from a sounder connected with a line employed in transmitting ordinary commercial and railway messages, in order that the student may familiarize himself with the technical usages of the lines, and the minute details of actual telegraphic business.[4]

380. **A Parting Word.**—In conclusion, the student is warned against falling into the common error, which is not confined to telegraphy, of expecting great results from little labor. To become an expert sending and receiving operator requires a vast amount of time and patience, and the most unwearied application. Remember that *whatever is worth doing at all, is worth doing well.* It is seldom that a thoroughly competent operator cannot obtain immediate and remunerative employment, and it is probable that such will continue to be the case, however crowded the lower walks of the avocation may hereafter become.

---

[4] Full explanations respecting the methods, regulations, and forms usually employed in the commercial, railway, and express service, in the forwarding and reception of messages, train orders, reports, etc., and much other miscellaneous information of like character useful to the student of telegraphy, may be found in the later editions of Abernethy's *Modern Service of Commercial and Railway Telegraphy.* A little work by T. J. Smith, on *The Philosophy and Practice of Morse Telegraphy*, may also be consulted with advantage.

INDEX.

ABERNETHY'S *Commercial and Railway Telegraphy* referred to, 223.
Absolute system of measurement, 37; concrete example of, 37.
Accumulator or storage battery, how shown in diagram, 104.
Accumulation, electrostatic, upon insulated conductor, 177.
Adjustment of key, 140; of quadruplex apparatus, 188; of register, 148; of sounder, 142.
Air, non-conducting properties of, 57.
Alloys of metals, inferior conducting power of, 57.
Alternating current, 167; rectification of, 167.
Amalgamation of zinc, 21.
American modification of closed-circuit system, 108.
American lines, defective insulation of, 118.
American electrical society, journal of, reference to, 187.
American institute of electrical engineers, extract from transactions of, 2.
American standard wire-gauge, 95.
Ammeter or amperemeter, the, 44, 61; use of in telegraphic testing, 210; Weston's portable, 211; advantages of for testing, 211.
Ammeter, Weston's combined voltmeter and, for telegraphic testing, 210.
Ampère, Andre Marie, biographical notice of, 60.
Ampère, the unit of current, definition of, 60; value of, 60; determination of, 60.
Ampère-turns, 85; magnetization proportional to, 87.
Ampèremeter or ammeter, the, 61.
Anderson's machine for winding helices of electro-magnets, 93.
Anthony, Wm. A., *Review of Modern Electrical Theories*, reference to, 2.
Apparatus, electric, drawings of, 103; telegraphic, conventional representations of, 103, 104.
Apparent resistance of line, 128; table of, 129.
Armature of magnet, 26; of electro-magnet, 91; polarized, 100.
Armature time of telegraph magnet determined, 99.
Artificial line of multiple telegraph, the, 174.
Artificial magnet defined, 24.
Astatic system of needles, 198; galvanometer, 200.
Attraction and repulsion, magnetic, 28; mutual, between electric conductors, 35.
Attraction, magnetic, ratio of to distance, 89.
Authors referred to:
  Abernethy, J. P., 223.
  Anthony, William A., 2.
  Avery, Elroy M., 70.
  Becker, C., 62.

Benoit, René, 57.
Bidwell, Shelford, 26.
Blavier, E. E., 17, 128.
Bonsanquet, R. H. M., 81.
Bottone, Selino R., 46.
Bradley, Leverett, 17, 76.
Brooks, David, 119, 120.
Cavendish, Henry, 2.
Chaperon, G., 23.
Christie, Samuel Hunter, 195.
Clark, Latimer, 58, 76, 125, 198.
Clerk-Maxwell, James, 2.
Cooke, Josiah P., 6.
Daniell, Alfred, 2, 38.
Davis, Daniel, Jr., 89.
Dean, G. W., 99.
Everett, J. D., 38.
Ewing, J. A., 98.
Faraday, Michael, 2, 3, 29, 38, 73, 88.
Farmer, Moses G., 62, 119, 128, 136.
Franklin, Benjamin, 2.
French, E. L., 87.
Gavarret, J., 128.
Gee, W. W. Haldane, 62.
Gray, Andrew, 40.
Grove, Sir William, 38.
Healy, Clarence I., 187.
Helmholtz, Herman L. F., 38.
Henning, Thomas, 187.
Henry, Joseph, 2, 80.
Hering, Carl, 82.
Hill, Edward A., 17.
Hughes, David E., 24.
Jamieson, Andrew, 73.
Jenkin, Fleeming, 17, 18, 62, 198.
Johnson, A. J., 20.
Jones, Francis W., 187.
Kapp, Gisbert, 24, 83.
Kempe, A. B., 128.
Kempe, H. R., 128, 198.
Kennelly, A. E., 85.
Kohlrausch, F., 128.
Lalande, de, F., 23.
Lockwood, Thomas D., 23, 34.
Lodge, Oliver J., 2, 88.
Maver, William, Jr., 185, 187.
Mayer, Alfred M., 24.
Mayer, Julius R., 38.
Morse, Samuel F. B., 128.
Munroe, John, 73.
Niaudet, Alfred, 62.
Nipher, Francis E., 40.
Nystrom, John W., 28.
Plum, H. W., 187.
Pope, Franklin L., 17, 76, 80, 187.
Preece, William H., 79.
Prescott, George B., 62.
Prescott, George B., Jr., 74, 94, 112.
Rowland, H. A., 81, 88.
Sabine, Robert, 83.
Schott, C. A., 40.

225

Shaffner, Tal P., 62.
Smith, T. Jarrard, 223.
Sprague, John T., 38, 57, 62, 74.
Stewart, Balfour, 38, 62.
Sturgeon, William, 85.
Thompson, Silvanus P., 31, 86, 87, 93, 198.
Thomson, Sir William, 2, 73, 88, 93.
Trowbridge, John, 40, 83.
Tyndall, John, 38.
Varley, Cromwell F., 131, 135.
Webb, F. C., 175.
Wheatstone, Sir Charles, 195.
Wilkinson, H. D., 85.
Youmans, Edward L., 38.
Automatic repeaters, management of, 166; Milliken's, 165.

BAD JOINT ON LINE, method of locating, 205.
Balancing of resistance in multiple telegraph, 174.
Bar magnet, 24.
Batteries, composed of number of cells, 3.
Battery, method of determining cost of maintenance of, 75; position of in closed-circuit system of telegraphy, 109; potentials within, 123; best position for on leaky line, 131, 132; internal resistance of, methods of measuring, 207.
Battery materials, choice of, 20.
Battery solutions, table of specific gravities of, 9.
Baume's hydrometer scale, 7.
Becker's experiments on resistances of liquids, 62, 63.
Benoit on specific resistance of metals, reference to, 57.
Bichromate of potash cell, 23.
Bidwell, Shelford, on maximum magnetic attraction, 26.
Binding screws, different patterns of, 50.
Biographical notices:
  Ampère, Andre Marie, 60.
  Coulomb, Charles Augustin de, 61.
  Faraday, Michael, 73.
  Gauss, Karl Friedrich, 83.
  Henry, Joseph, 80.
  Joule, James Prescott, 63.
  Lefferts, Marshall, 214.
  Morse, Samuel Finley Breese, vi.
  Ohm, Georg Simon, 62.
  Vail, Alfred, 216.
  Volta, Alessandro, 61.
  Watt, James, 73.
Blavier's *Télégraphie Electrique*, reference to, 17, 128.
Bonsanquet, R. H. M., on magneto-motive force, reference to, 81.
Boston screw-glass insulator, tests of efficiency of, 120.
Bottone's *Electrical Instrument-making for Amateurs*, extracts from, 46.
Box sounder, 144; use of in railway service, 144.
Bradley, L., on Hill's gravity cell, 17.
Branch circuit connection, diagram of, 104.
Branch or derived circuits, 69; rule for joint resistance of, 66.
Brass, specific resistance of, 57.
Break or disconnection, conditions arising from, 190.
Breakage of battery jars, causes of, 10.
Bridge, Wheatstone's, 195; theoretical arrangement of, 195; invented by Christie, 195; principle of illustrated, 196; best ratio of electromotive forces and resistances in, 196; actual construction of, 198; galvanometer for, 198; methods of making various tests with, 199–208.

British association ohm, determination of, 62, 63; value of, 62.
British standard wire-gauge, 95.
Brooks, David, on effects of climate upon telegraphic insulation, 119; on effects of smoke in cities on insulation, 119; tests of various kinds of insulators by, 120.
Brooks's paraffin insulator, tests of efficiency of, 120.
Brown & Sharpe M'f'g Co.'s American standard wire-gauge, 95.
Brushes of dynamo-electric machine, 168.
Bunsen's nitric-acid cell, 23.
Button repeater, the, 162; management of, 165.

CABLE, submarine, diagram of, 104.
Caliper, micrometer, for gauging wires, 96.
Callaud's cell, 17.
Calorimeter, the, 44.
Canada, closed circuit used in, 108.
Capacity, inductive or electrostatic, 72; definition of, 72; unit of, 73.
Cavendish's theory of electricity, 2.
Cell, gravity, maintenance of, 14; dismantling of, 16; best adapted to closed circuits, 16; waste products of, 17; electromotive force of, 74; resistance of, 74.
Cell, usual internal resistance of, 64; electrical dimensions of, 74.
Cell, oxide of copper, 21.
Cell, sulphate of copper, effect of temperature upon resistance of, 78.
Cell, voltaic, Hill's, Callaud's, Minotto's, Thomson's, 17; Lockwood's, 18; Daniell's, 19; Edison-Lalande, 21; Grove's, 23; Bunsen's, 23; rate of consumption of material in, 13; effect of continued action in, 13; various forms of, 17; general directions for care of, 20; how shown in diagram, 104.
Centimetre, the unit of space, 37.
Centimetre-gram-second system of units, 37; units of force and work, 37.
Chaperon, G., and F. de Lalande, on voltaic batteries, 23.
Characters, code, exercises with, 220.
Charge, electrostatic, 177; current of, 177.
Chemical atomic weights of battery materials, table of, 8.
Chemical electricity, 3.
Chemical equivalents, table of, 75.
Chemical law of definite proportion, 6.
Chemical reaction of voltaic cell, 6; in closed circuit, 12.
Chemistry of voltaic effect, 6.
Circuit, conducting, effect of increasing the length of, 54.
Circuit, closed, the, 12; chemical reactions in, 12; theoretical diagram of, 12.
Circuit, distribution of potentials in when insulated, 120.
Circuit, electric, constituent parts of, 11; formation of, 11; nomenclature of, 12; graphic illustration of, 71.
Circuit, external, the, 12; internal, the, 12.
Circuit, imperfectly insulated, distribution of potentials in, 125.
Circuit, magnetic, 80; conception of due to Joule, 80.
Circuit, open or broken, the, 12.
Circuits, telegraphic, 102; open and closed, 102; diagram of, 104; essential characteristics of, 102; general considerations respecting, 109; working efficiency of, 111; distribution of potentials in, 120.
Circuits of American telegraph system, arrangement of, 151.

Index. 227

Clamp-screw of gravity cell, 4.
Clark, Latimer, on Wheatstone's bridge, 193; provisional theory of electricity, 58.
Clark's *Electrical Measurement*, references to, 125, 198; extract from, 58.
Clerk-Maxwell, theory of electricity, 2.
Climate, effect of upon insulation, 118.
Clip, the, in diplex and quadruplex telegraph, how obviated, 184.
Closed and open circuit systems of telegraphy, comparative advantages of, 109.
Closed-circuit system of telegraphy, 102; description of, 108; American modification of, 108; position of battery in, 109.
Coast survey report, reference to, 99.
Cobalt, magnetic properties of, 24.
Code, telegraphic, formation of, 216; elementary principles of, 219.
Code, American Morse, 217, 218; alphabet and numerals of, 218; punctuation, etc., of, 218; best method of learning, 219; exercises with, 220.
Coil or loose bundle of wire, how shown in diagram, 104.
Combinations of permanent and electro-magnets, 100.
Commutator of dynamo-electric machine, function of, 167; construction of, 168.
Compass, magnetic, 25.
Condenser, construction of, 178; application of to duplex telegraph, 178; first applied by Stearns, 178; how shown in diagram, 104.
Conducting circuit, an element of the electric telegraph, 2.
Conductivity resistance, relation of insulation to, 110; of line, measurement of with bridge, 199.
Conductors, insulated, for interior construction, 114; telegraphic, 111.
Conductors and insulators, characteristics of, 56; table of, 57.
Connecting wire in gravity battery, protection of, 11.
Conservation of force, principle of explained, 38.
Constancy, value of in voltaic cell, 74.
Constant of galvanometer, method of determining, 206.
Consumption of material in cell in relation to electricity evolved, 17.
Contraplex and diplex methods, combination of, 184.
Conventional representation of circuits and apparatus, 103, 104, 105.
Cooke's *New Chemistry*, extract from, 6.
Copper, chemical equivalent of, 75.
Copper connector, of gravity cell, 5.
Copper plate, of gravity cell, 4; modifications of, 10; the negative element of, 12.
Copper line wires, 114; table of dimensions and qualities of, 112.
Copper sulphate, chemical analysis of, 8.
Copper wire, bare, for magnet helices, 94; hard drawn, table of sizes, weights, resistances, etc., 112.
Core of electro-magnet, 91.
Core, diameter of in electro-magnet, force of attraction affected by, 87; best proportions for, 91.
Cost of materials consumed in battery, 74; of battery maintenance, method of determining, 75.
Coulomb, Charles Augustin de, biographical notice of, 61.
Coulomb, the unit of electrical quantity, definition of, 61.
Cross, definition of, 190; metallic, 191;
swing, 191; weather, 191; method of testing for, 193; principle of test for, 193.
Cross, on line, locating position of, 204.
Cross-arms, tests of insulating value of, 133, 134.
Cross-current, remedy for, 132.
Cross-fire, 191; explanation of cause of, 132; remedy for, 132.
Crossing of two wires, representation of in diagram, 104.
Cross-section of body, effect of upon resistance, 58.
Current, alternating, of dynamo-machine, 167.
Current, electric, formation of, 11; produced by magnetic field, 29; manifestations of in conductor, 35; effect of imperfect insulation upon flow of, 127; direction of, how shown in diagram, 104.
Current force, relation of to mechanical force, 40.
Current induction, 74.
Current, inducing or primary, 74; direction of, 31.
Current, induced or secondary, 31, 74.
Current in leaky lines, 128; table for computing, 129.
Current of charge on insulated line, 177.
Current of dynamo-electric machine, characteristics of, 167.
Current, relation of to magnetic force, 85; self-induction of, 98; in coiled conductor, 98.
Currents, adaptation of electro-magnets to, 96; method of determining, 96; distribution of in quadruplex telegraph, 186.
Currents, earth, disturbing influence of on conductivity tests, 201.
Currents of charge and discharge, effects of on line, 177.
Currents, received, test of insulation by means of, 208.
Curve, of ratio between magnetic attraction and distance, 90; of electrical dimensions in oxide of copper cell, 77; of resistance as affected by temperature in Daniell's cell, 79; of magnetization of soft iron, 85; of magnetic saturation, 85; of potentials in electric circuits, 121, 123, 124; of potential within battery, 125; of potential on leaky line, 125, 126, 130.
Cut-out wedge, the, 155.

DANIELL's *Principles of Physics*, extract from, 2; reference to, 38.
Daniell's sulphate of copper cell, 18; maintenance of, 19; renewal of, 19.
Davis, Daniel, Jr., experiments on magnetic attraction. 89, 90.
Davis's *Manual of Magnetism*, reference to, 89.
Dead ground, definition of, 190.
Dean, G. W., experiments of on self-induction and hysteresis in telegraph magnets, 99.
Deflections, proportional, measuring high resistances by method of, 205.
De Lalande, F., and G. Chaperon, on voltaic batteries, 21.
Density, magnetic, 83.
Derived and fundamental units, 37.
Derived or branch circuits, 69.
Detector or galvanoscope, 41.
Diagrams of electric apparatus, 103.
Differential, electro-magnet, principle of, 171; galvanometer, construction and use of, 208.
Diffusion of solution in gravity cell, 16.

Dimensions, electrical, of voltaic cell, 74.
Diplex telegraphy, 171.
Diplex, principle of, 182; receiving apparatus of, 183; clip in, 184; short core relay for, 184.
Diplex and contraplex, combination of, to form quadruplex, 184.
Direction of electric current, purely a conventional assumption, 12.
Disconnection or break, conditions arising from, 190; testing for, 191; testing for at way station, 154; causes of, 192.
Disconnection, partial, 190; testing for, 192.
Distance between magnet and armature, effect of upon attractive force, 89; experimental determination of, and tabulated results, 89, 90.
Dot, an element of telegraphic code, 216; the unit of time and space in ditto, 217.
Double current duplex, apparatus of, 180.
Double current or reversing key, 180; how shown in diagram, 104.
Double measurement, process of in line testing, 203.
Drawings, of electric apparatus, 103; perspective, 103; geometrical, 103.
Duplex telegraphy, 171.
Duplex, single current, 172; apparatus of, 172; circuits of, 173; artificial line of, 173; balancing of, 173; effect of currents of charge and discharge in, 177; ground and spark coils of, 179; double current, description of, 180.
Duration of cell, considerations affecting, 74.
Dynamo current, characteristics of, 167; application of to quadruplex telegraphy, 185, 186.
Dynamo-electricity defined, 24.
Dynamo-electric generator, employment of in telegraphy, 167.
Dynamo-electric machine, the, 32; theory of explained, 32; diagram of, 104; field of, 168; commutator of, 168; brushes of, 168; characteristics of, 169; Edison's, 168; arrangement of in potential series, 169; positive and negative series of, 170; capacity of, 171; shunt coils of, arrangement of in telegraphy, 171.
Dynamos, arrangement of in series in New York station, 170.
Dyne, unit of force, definition of, 38.

EARTH, the, an electrical conductor, 104; magnetism of, 40; field of force due to, 40.
Earth circuit, principle of, 106; advantages of, 106.
Earth or ground plate, how shown in diagram, 104; precautions in fixing, 106.
Earth currents, disturbing influence of on conductivity tests, 201; how eliminated, 201.
Edison, T. A., inventor of method of diplex transmission, 185.
Edison's dynamo-electric machine, 168, 169.
Edison-Lalande oxide of copper cell, 21; electromotive force and resistance of, 76; duration of, 77; chart of electrical dimensions of, 77.
Effect of continued action on voltaic cell, 13.
Efficiency, working, of lines, importance of high, 135; best method of improving, 135; examples of advantageous results of, 135, 136; of telegraphic circuit, 111; computation of, 128.
Electric circuit, formation of, 11; graphic illustration of, 71.
Electric current, manifestations of in conductor, 35; produced by magnetic field, 29.

Electric field of force, 39.
Electrical action, laws and conditions of, 45; mechanical analogue of, 58.
Electrical and magnetic units, derivation of, 37.
Electrical and mechanical force, statement of law connecting, 59.
Electrical Engineer (N. Y.), references to, 23, 74; extract from, 2.
Electrical measurement, quantitative, theory of, 35; importance of, 36.
Electrical Review (London), reference to, 23.
Electrical World, reference to, 185.
Electrician (London), reference to, 81.
Electrician and Electrical Engineer, references to, 87, 94.
Electricity, chemical, 3; magneto, 24; dynamo, 24; frictional, 33; static, 33; thermo, 34.
Electricity, theories of nature of, 2; origin of, 3; sources of, 3; characteristics of capable of measurement, 43; apparatus required for measurement of, 43; provisional theory of, 58; production of in battery in proportion to material consumed, 76.
Electricity and magnetism, essential nature of, 2.
Electrification, 72.
Electro and permanent magnets, combinations of, 101.
Electro-chemical equivalent, of zinc, 74, 75; of copper, 75; of copper sulphate, 75.
Electrodes of cell defined, 12.
Electrolysis of liquids by electric current, 36.
Electro-magnet, the, 80; its modern form invented by Henry, 80; polarity of determined by direction of current, 81; elements of, 81; adaptation of to working currents, 96; spectrum of, 96, 97; indirect causes of retardation in, 99; with polarized armature, 100; differential, principle of, 171; construction of, 172.
Electro-magnetism, 32; laws of, 80.
Electromotive force, conception of, 59; of ordinary gravity cell, 74.
Electrostatic or inductive capacity, 72; of line, 175.
Electrostatic accumulation upon insulated conductor, 177.
Electrostatic balance of duplex telegraph, 178.
Elements of electric telegraph, 2; of electro-magnet, 81.
Endo-mose, action of in voltaic cells, 20.
English unit of magnetic induction, 83.
Equator of magnet, 26.
Equipment of American telegraph lines, 138.
Equivalent, of mechanical energy, 31; electrical and mechanical, definition of, 38.
Erg, unit of work, definition of, 38.
Escape or leakage on line, 190; testing for, 192; in line, locating position of, 202; ditto by double measurement, 203; by loop test, 203.
European register, 148.
Evaporation in battery cells, prevention of, 10. 14.
Everett's *Units and Physical Constants*, reference to, 38.
Ewing's researches in magnetism, 98.
Exercises with code characters, 220, 221.
External circuit, the, defined, 12.

FALL OF POTENTIAL, illustrated, 70; proportionate to fall of resistance along conductor, 71.
Farad, the unit of electrostatic or inductive capacity, 73.

Index. 229

Farad.y, Michael, biographical notice of, 73.
Faraday's theory of electricity, 2 ; discovery of magneto-electricity, 29 ; *Experimental Researches*, references to, 3, 29 ; lines of magnetic force, 27, 28.
Farmer, M. G., on resistances of battery solutions, 62 ; observations on earth circuit as affected by character of soil, 107 ; on effects of wet upon telegraphic insulation, 119 ; on working efficiency of telegraph lines, 128; table of percentages of received current on telegraph lines, 136.
Fault in line, effect of position of, 131.
Faults and interruptions, classification of, 190.
Field, electro-magnetic, of dynamo machine, 168.
Field, magnetic, the, 27.
Field, magnetic, lines of force a measure of, 82.
Field of force, electric, 39.
Field of magnetic force, properties of, how determined, 27.
Field, Stephen D., inventor of application of dynamo to telegraphy, 185.
Flux, magnetic, 88.
Force, conservation of, doctrine of, explained, 38.
Force, definition of, 27 ; lines of, a measure of magnetic field, 82 ; relation of current to mechanical, 40 ; unit of, defined, 37.
Formation of the electric circuit, 11.
Franklin's theory of electricity, 2.
French, E. L., on electro-magnetic attraction, 87.
Frictional electricity, 33.
Fundamental and derived units, 37.
Fundamental units of mass, space and time, 36.

GALVANIC ELEMENT, the, 3.
Galvanized iron wire, table of sizes, weights and resistances of, 112.
Galvanometer, the, 41 ; how shown in diagram, 104 ; astatic, 200 ; differential, construction of, 208 ; for Wheatstone bridge, 198.
Galvanometer, tangent, 41, 209 ; construction of, 41 ; use of in testing insulation by received currents, 209 ; in experimental investigations, 52 ; table of tangents for, 55.
Galvanometer, taking constant of, 206 ; resistance of, method of measuring, 208 ; application of shunts to, 205.
Galvanoscope, the, 41 ; how shown in diagram, 104.
Gap in magnetic circuit, effect of, 89.
Gas and water pipes used for ground connections, 106.
Gauging wire, instruments for, 95.
Gauss, Karl Friedrich, biographical notice of, 83.
Gauss, the proposed unit of magnetism, 83 ; definition of, 83.
Gauss and Weber's electric telegraph, 83.
Gavarret's *Telegraphie Électrique*, reference to, 128.
Generator, the, an element of electric telegraph, 2.
German-silver, specific resistance of, 57.
Glass insulator, the, 115; defects of, 115; tests of, 120, 134.
Glass jar of battery cell, 4 ; breakage of, how avoided, 10.
Gram, the unit of mass or weight, 37.
Gravity cell, description of, 4 ; installation of, 9 ; formula for preparing solutions for, 8.

Gray's *Absolute Measurements in Electricity and Magnetism*, reference to, 40.
Ground or earth plate, how shown in diagram, 104 ; at dis.ant station, measuring resistance of, 202 ; defective, effect of, 191.
Ground connection, how made, 106 ; precautions in making, 106.
Ground, on line, 190 ; locating position of, 202.
Ground and spark coils in duplex telegraph, 179.
Grove's nitric acid cell, 23.

HANGER OF GRAVITY BATTERY CELL, 4.
Hardening of magnet cores objectionable, 99.
Hard iron and steel, magnetic properties of, 24.
Hard-rubber insulator, the, 117.
Healy, Clarence L., on quadruplex telegraphy, 187.
Heat, development of by electric current, 36 ; effect of upon *e. m. f.* of sulphate of copper cell. 74.
Helices of electro-magnets, machine for winding, 93 ; thickness of spaces between wires of, 94 ; of bare copper wire, 94.
Helix, magnetic, effect of iron in, 84 ; effect of position of windings in, 92 ; construction of, 92 ; relation of number of turns to thickness and length of wire in, 92 ; number of turns in, measured by its resistance, 93.
Henning, Thomas, on quadruplex telegraphy, 187.
Henry, Joseph, biographical notice of, 80.
Henry's theory of electricity, 2.
Hering, Carl, on lines of magnetic force, 82.
Hering's *Principles of Dynamo-Electric Machines*, extract from, 82.
High resistances, methods of measuring, 205.
Hill, E. A., on the voltaic cell, 17.
Hill's cell, 17.
Hints to learners, 216.
Horizontal component of earth's magnetism, defined, 40 ; value of in various parts of North America, 40.
Horseshoe, magnet, 26 ; electro-magnet, 87.
Hughes, Prof. D. E., theory of magnetism, 24.
Hydrogen, evolution of in voltaic cell, 5.
Hydrometer, the, description of, 7.
Hysteresis, magnetic, definition of, 97; effect of on telegraph magnets, 99.

INDUCED OR SECONDARY CURRENT, 74 ; direction of, 31.
Inducing of primary current, 74.
Induction, current, 74 ; static, 175, 177; magnetic, phenomena of, 25 ; magnetic, cause of, 85.
Induction of current upon itself, 98.
Inductive or electrostatic capacity, 72.
Inertia, magnetic, definition of, 99.
Ink-writing register, 150.
Installation of gravity cell, 9.
Instrument tables, 160.
Insulating value of wet poles and cross-arms, tests of, 133, 134.
Insulation, imperfect, effects of, 110 ; defective of American lines, 118 ; effects of climate upon, 118 ; effect of upon flow of current, 127 ; Farmer's table of, 136.
Insulation resistance of line, measurement of, 202 ; relation of conductivity to, 110.
Insulation, test of, by received currents, 208.
Insulator, glass, Western Union, old pattern, 116 ; new standard pattern, 116 ; hard rubber, 117 ; paraffin, 117 ; porcelain, 118.

Index.

Insulators and conductors, characteristics of, 56; comparative table of, 57.
Insulators, telegraphic line, 115; common glass, 115; defects of, 115; resistance of influenced by form, 116; comparison of different forms of, 116; tests of, 120, 134; measurement of resistance of, 206; value of w-t poles and cross-arms considered as, 133.
Intensity, of magnetization defined, 86; of magnetic field defined, 27.
Intermingling of currents on telegraph lines, 132.
Internal circuit, the, defined, 12.
Internal resistance of battery, methods of measuring, 207.
Interruptions and faults on telegraph lines, classification of, 190.
Iron, magnetization of by electric current, 35; specific electrical resistance of, 57; effect of in magnetic helix, 84.
Iron filings held to conductor by magnetism, 35.
Iron wires for lines, table of sizes, weights, and resistances of, 112; joints in, 113.
Iron and steel, magnetic properties of, 24.

JENKIN, FLEEMING, on conductors and insulators, 56; on Wheatstone's bridge, 198.
Jenkin's *Electricity and Magnetism*, references to, 17, 18, 62, 198.
Johnson's *Universal Cyclopedia*, reference to, 20.
Joint, defective, on line, method of locating, 205.
Joint, twist, for iron wires, 113; importance of soldering, 113.
Joint resistance of branch circuits, rule for determining, 66.
Jones, Francis W., on quadruplex telegraphy, reference to, 187.
Joule, James Prescott, biographical notice of, 63.
Joule's law, statement of, 63.
Joule's conception of magnetic circuit, 81.

KAPP LINE, the, 83.
Kapp's *Electric Transmission of Energy*, reference to, 24.
Kapp's unit of magnetic induction, 83; value of, 83.
Kempe, A. B., on the leakage of submarine cables, reference to, 128.
Kempe, H. R., *Handbook of Electrical Testing*, references to, 128, 198; on Wheatstone's bridge, 198.
Kennelly and Wilkinson's *Practical Notes for Electric Students*, reference to, 85.
Kerite insulation for office wires, 113, 114.
Key, construction of, 138; adjustment of, 140; platinum contacts of, 138; modifications of, 139; Western Electric pattern, 140; Victor pattern, 140; double-current or reversing, 180; common Morse, how shown in diagram, 104; three-point, how shown in diagram, 104; method of handling, 219; preliminary practice with, 220.
Key and sounder, combination, 143; pocket, 144.
Key, relay, and register, combination of, 149.
Kick, due to electrostatic charge and discharge, 177.
Kohlrausch's *Physical Measurement*, reference to, 62.

LAW OF MAGNETIC CIRCUIT, 88.
Law of electric current, Ohm's, 63; Joule's, 63.

Law connecting mechanical and electric force, 59.
Learners, hints to, 216.
L'Electrician, reference to, 23.
Lefferts, Marshall, biographical notice of, 214.
Legal ohm, determination of, 63; value of, 63.
Length of body, effect of upon resistance, 58.
Letter space, an element of telegraphic code, 216.
Lightning arrester, description of, 160; combination of with switch, 156, 157; inspection and care of, 161; how shown in diagram, 104.
Line, computation of working efficiency of, 128; electrostatic capacity of, 175; overhead, diagram of, 104; submarine or subterranean, diagram of, 104; artificial, of multiple telegraph, 174.
Lines, leaky, resistance and current in, 128; table for computing resistances and escapes upon various lengths of, 129; best position for battery on, 131, 132; effect of pos.tion of fault in, 131.
Lines of magnetic force, conception of due to Faraday, 27; rendered visible by iron filings, 27.
Local circuit, working by relay and, 144.
Lockwood cell, description of, 18.
Lockwood's *Electricity, Magnetism, and Electric Telegraphy*, references to, 23, 34.
Lodestone or natural magnet, properties of, 24.
Lodge, Oliver J., theory of electricity, 2.
Loop test for conductivity resistance, 201; for position of escape, 203; Varley's modification of, 203.

MAGNET, bar, 24; horseshoe, 26; multipolar, 26; natural and artificial, 24; temporary and permanent, 24.
Magnet, polarity of, 25; magnetic length of, 26; maximum attraction exerted by, 26.
Magnet, telegraph, best proportions for, 91; details of construction of, 91.
Magnet wires, Prescott's table of dimensions and resistances of, 93.
Magnetic attraction and repulsion, 28.
Magnetic circuit, conception of by Joule and others, 81; formation of, 87; law of, 88.
Magnetic compass, the, 25.
Magnetic density, 83.
Magnetic field, the, 27; exploration of by suspended needle, 28; of the earth, 40.
Magnetic flux, 88.
Magnetic force, relation of current to, 85.
Magnetic hysteresis, definition of, 97.
Magnetic induction, phenomena of, 25; cause of, 85.
Magnetic inertia, definition of, 99.
Magnetic length of magnet, 26.
Magnetic meridian, the, 24, 28.
Magnetic moment, definition of, 86; experimental determination of, 86; tabulated results of, 86.
Magnetic needle, the, 24.
Magnetic permeability, 88; definition of, 88.
Magnetic reluctance, 88; determination of, 88.
Magnetic resistance, 88.
Magnetic saturation, curve of, 85; definition of, 87.
Magnetic spectrum, the, 26.
Magnetic and electrical units, derivation of, 37.
Magnetism, definition of, 24; characteristics of, 24; unit of, 83; intensity of, 83; density of, 83; lines of force a measure of, 82; remanent or residual, 98.

Magnetism and electricity, essential nature of, 2.
Magnetization, of one body by another, 25; intensity of defined, 86; proportional to ampere-turns, 87; maximum limit of in soft iron, 87.
Magneto-electricity, 3; definition of, 24; phenomena of, summarized, 32.
Magneto-motive force, 81, 83; definition of, 83; method of computing, 84.
Magnetometer, construction and use of, 85.
Maintenance of voltaic cell, 14.
Manual and automatic repeaters, 162.
Manuscript copy, practice from recommended, 222.
Materials consumed in battery, quantity and cost of, 74.
Matthiessen on specific resistance of metals, reference to, 57.
Maver, William, Jr., on dynamo telegraphy, 185; on quadruplex telegraphy, 187.
Maximum magnetic effect in electro-magnet, best proportions for, 91.
Mayer's *Earth a Great Magnet*, reference to, 24.
Measurement, absolute system of, 37; quantitative electrical, theory of, 35; importance of, 36; electrical, character of, 43; practice of, 195.
Mechanical force, relation of to current force, 40.
Mechanical power, transformation of into electricity and heat, 31.
Mechanical analogue of electrical action, 58.
Megadyne, definition of, 38.
Metallic cross, 191.
Metals, specific resistance of, 57.
Metric system, foundation of absolute system of electrical and magnetic units, 37.
Mexico, closed circuit used in, 108.
Microfarad, the, 73.
Micrometer caliper for gauging wires, 96.
Milliken's automatic repeater, 165.
Minotto's cell, 17.
Modern Practice of Electric Telegraph (4th ed.), reference to, 17, 23.
Moment, magnetic, definition of, 86.
Movement of conductor in magnetic field, effects of, 30.
Muller, on increase of resistance of metals by rise in temperature, 78.
Multiple series of cells, 53.
Multiple wire switchboard, 156.
Multiples of electrical units, 60.
Multipolar magnet, 26.
Munroe and Jamieson's *Pocket-book of Rules and Tables*, reference to, 73.
Mutual reactions of current and magnet, 32.

NEEDLE, magnetic, deflected by current, 35.
Needles, astatic, system of, 198.
Negative plate or element of cell, 12.
Neutral line of magnet, 26.
Niaudet's *Electric Batteries*, reference to, 62.
Nickel, magnetic properties of, 24.
Nipher's *Theory of Magnetic Measurements*, reference to, 40.
Nomenclature of electric circuit, 12.
North pole and south pole of magnet, 26.
North-seeking pole, the, 29.
North British Review, extract from, 56.
Nystrom, J. W., definition of force, 27.
Nystrom's *Elements of Mechanics*, extract from, 27.

OFFICE WIRES, 113; table of dimensions of, 114.
Ohm, Georg Simon, biographical notice of, 61.

Ohm, the unit of electrical resistance, definition of, 61; value of, 61, 62, 63.
Ohm's law, statement of, 63; experimental proof of, 64.
Oil, used to prevent evaporation in voltaic cells, 14.
Okonite insulation for office wires, 113, 114.
Open circuit, in telegraphy, 102, 103; system of telegraphy, description of, 107.
Open and closed circuit, comparative advantages of, 109.
Operator and Electrical World, references to, 185, 187.
Origin of electricity. 3.
Oxide of copper cell, 21; maintenance of, 22; chemical reaction, 22.
Oxide of zinc formed by action of voltaic cell, 6.

PARAFFIN INSULATOR, the, 117.
Parallel arrangement of voltaic cells, 54.
Parallel series of cells, 53.
Paris exposition, *Report on Telegraphic Apparatus*, extract from, 128.
Parting word, a, 223.
Permanent and temporary magnets, 24.
Permanent and electro-magnets, combinations of, 101.
Permeability, magnetic, definition of, 88; of iron, 81.
Phenomena, of voltaic cell, 5; of induction upon telegraph lines, 177.
Philosophical Magazine, reference to, 81.
Philosophical Transactions of Royal Society, references to, 26, 98.
Physiological effects of electric current, 36.
Platinoid, specific resistance of, 57.
Platinum, specific resistance of, 57.
Plum, H. W., on quadruplex telegraphy, 187.
Pocket telegraphic apparatus, 144.
Polar or polarized relay, 180; how shown in diagram, 104.
Polarity of the magnet, 25; of electro-magnet determined by direction of current, 81.
Polarized armature, 100.
Pole changer, peg, construction of, 51.
Pole-changing transmitter in multiple telegraphy, 181.
Poles of magnet, 25; of voltaic cell, 12.
Poles, tests of insulating value of when wet, 133, 134.
Pope, F. L., on quadruplex telegraphy, 187.
Porcelain insulator, the, 118; tests of, 120.
Porous cell, action of, 20.
Position, method of locating, of ground, 202; of escape, 202; of cross, 204; of bad joint or abnormal resistance, 205.
Positive plate or element of cell, 12.
Post-office, British, table of iron wires, 112.
Potential, conception of, 59; of telegraph line, determination of by calculation, 123; of line, measurement of by auxiliary battery, 122.
Potential, electric, explanation of, 69; fall of along conductor, illustration of, 70; proportionate to resistance, 71; in perfectly insulated circuit, distribution of, 120; in imperfectly insulated circuit, 125; within battery, 123.
Potential series, arrangement of dynamos in, in telegraphy, 169.
Power or rate of work of electric current, 73.
Practical electric units, 59.
Practice, preliminary, with key, 220.
Preece, W. H., on effects of temperature on sulphate of copper cell, 78, 79.
Prescott, Geo. B., Jr., on electro-chemical

equivalents, 74; table of properties of soft copper wires, 112.
Prescott's *Electricity and Electric Telegraph*, reference to, 62.
Primary or inducing current, 74.
Properties and dimensions of copper magnet wires. Prescott's table of, 94.
Proportional deflections, measuring high resistances by method of, 205, 206; measuring resistance of insulators by, 206.

QUADRUPLEX TELEGRAPHY, 171; principle of, 182; a combination of diplex and contraplex systems, 184; distribution of currents in, 186; how worked by dynamo currents, 185; practical management of, 187; adjustment of apparatus of, 188; repeaters for, 189.
Quantity, of current in circuit, conditions which determine, 54; of material consumed in battery, 74.

RAIN WATER, should be used in batteries, 5.
Rate, of consumption of material in voltaic cell, 13; of work of electric current, how found, 64; relation of to time, 73.
Reading by sound, 222.
Receiver, the, an element of the electric telegraph, 2.
Reciprocals, 66; table of, 67.
Reduced length of conductor, meaning of, 58.
Register, the, 147; construction of, 147; European pattern of, 148; combination of with key and relay, 149; adjustments of, 149; causes of defective marking in, 150; ink-writing, 150; how shown in diagram, 104.
Relay, construction of, 144; function of, 146; adjustments of, 147; short core, for diplex and quadruplex apparatus, 184; polar, or polarized, 180; how shown in diagram, 104; common or non-polarized, how shown in diagram, 104.
Relay and local circuit, working by, 144.
Relay, key, and register, combination of, 149.
Reluctance, magnetic, 88; determination of, 88.
Remanent or residual magnetism, 98.
Repeater, the, 162; manual and automatic, 162; button, 162; Wood's, 164; Milliken's, 165; for duplex, quadruplex, and multiple systems, 189.
Residual or remanent magnetism, 98.
Resistance, electrical, explanation of, 56; conditions affecting, 58; expressible in terms of length, 58; artificial, how shown in diagram, 104.
Resistance, of circuit, relation of to quantity of current flowing in, 56; abnormal on line, method of locating, 205; relation of conductivity to insulation of line, 110; ratio of conductivity to insulation, minimum, 128.
Resistance, conductivity, of line, measurement of with bridge, 199.
Resistance, insulation, of line, measurement of, 202; of insulators, method of measuring, 206; of various kinds of insulators, 120.
Resistance, joint, law of, 66; of a circuit, how to determine, 66.
Resistance, specific, of different metals, 57; of copper wires for electro-magnets, table of, 94; of galvanized iron, hard-drawn and soft copper wires, Prescott's table of, 112; of metals employed as conductors in telegraphy, percentage of increase in by rise of temperature, 78.

Resistance of liquids, Farmer's values of, 62; Becker's ditto, 63; of sulphate of copper solution, 62, 63; of sulphate of zinc solution, 62, 63; internal, of voltaic cell, 64; of ordinary gravity cell, 74; of sulphate of copper cell, effect of temperature upon, 78; experiments of Preece on, 78; internal, of battery, methods of measuring, 207.
Resistance, very high, methods of measuring, 205; of galvanometers, method of measuring, 208; of helix, a measure of number of turns in, 93; of leaky lines, 128; table of, 129.
Resistance coils, construction of, 51.
Re istance, magnetic, 88.
Retardation in electro-magnets, indirect causes of, 99.
Reversing or double-current key, 180.
Rheostat or artificial resistance, construction of, 51; how shown in diagram, 104.
Ring gauge, for wire, 95.
Roebling, J. A., table of hard-drawn copper wires, 112.
Rowland, H. A., *On Magnetic Permeability of Iron*, 81.
Royal Society's Proceedings, references to, 24, 79.

SABINE'S *Electric Telegraph*, reference to, 83.
Safety-fuse, use of, 161.
Saturation, magnetic, definition of, 87; curve of, 85; experimental investigation of, 86.
Schott, C. A., on determination of value of horizontal component of earth's magnetism, 40.
Second, the unit of time, 37.
Secondary or induced current, 74.
Self-exciting dynamo-electric machine, 163.
Self-induction, effect of on telegraph magnets, 99.
Sentence space, an element of telegraphic code, 216.
Series arrangement of voltaic cells, 52.
Series of cells, effect of varying number in, 53; consumption of material in, 76.
Shaffner's *Telegraph Manual*, reference to, 62.
Short-circuiting a cell, 8.
Short-line sounder, 142.
Shunt, definition and derivation of, 69.
Shunts, applied to galvanometer for measurement of high resistances, 205; method of using, 206.
Shunt-wound dynamo-electric machine, 168, 169.
Siemens' mercury unit, value of, 62.
Single-current duplex, the, 172.
Sizes of galvanized iron and copper wires, table of, 112.
Soft iron, effect of magnetization upon, 85; maximum limit of magnetization in, 87.
Soil, earth circuit affected by characteristics of, 106, 107.
Sound, reading by, 222; methods of practice in, 223.
Sounder, the, 141; adaptation of to short lines, 142; adjustment of, 142; combination key and, 143; pocket, 144; box, 144; how shown in diagram, 104.
Sources of electricity, 3, 24.
Space, an element of telegraphic code, 216.
Spaced letters, formation of, 222.
Spaces, in helices of electro-magnets, thickness of, 94.
Spark and ground coils in duplex telegraph, 179.
Specific gravity, explanation of, 7; of battery solutions, table of, 9.

Index. 233

Specific resistance of materials, 57; how determined, 57.
Spectrum, magnetic, the, 26; of electro-magnet, 96.
Spelter for battery zincs, analysis of, 20.
Sprague, J. T., on effect of heat on $e. m. f.$ of sulphate of copper cell, 74.
Sprague's *Electricity*, extract from, 74; references to, 38, 57, 62.
Spring-jack and wedge cut-out, the, 155; multiple ditto, 156.
Static or frictional electricity, 33.
Static charge of line, nature of, 177.
Stearns, Joseph B., his inventions in duplex telegraphy, 178.
Stewart and Gee's *Elementary Practical Physics*, reference to, 62.
Stewart's *Conservation of Energy*, reference to, 38.
Storage battery or accumulator, how shown in diagram, 104.
Student, apparatus required by, 45.
Sturgeon's Annals, reference to, 81.
Sub-multiples of electrical units, 60.
Sulphate of copper, chemical analysis of, 8; chemical equivalent of, 75; solution, resistance of, 62, 63.
Sulphate of zinc, chemical analysis of, 9; solution, resistance of, 62, 63.
Swinging cross, 191.
Switch, universal, how shown in diagram, 104; three-point, how shown in diagram, 104; pole-changing, how shown in diagram, 104.
Switchboard, multiple wire, 156; universal, 156; terminal, 158.
Switchboard for way-station, description of, 152; manipulation of, 153; testing by means of, 154.

TANGENT GALVANOMETER, description of, 41; principle of, 42; details of construction of, 45; Western Union pattern, 209; use of in testing insulation by received currents, 209.
Tangents, natural, table of, 55.
Tapping upon key, importance of avoiding, 219.
Tap-wire, in quadruplex telegraph, 187.
Telegraph, electric, elements of, 2.
Telegraph lines, mutual current induction between, 74; American, equipment of, 138.
Telegrapher, The, extracts from, 119, 120, 136; references to, 17, 187.
Telegraphic circuits, 102; distribution of potentials in, 120.
Telegraphic code, formation of, 216; elements of, 216; a multiple of time, 216.
Telegraphic conductors, 111.
Telegraphic magnet, theoretical proportions of, 91; details of construction of, 91; spectrum of, 97; effects of self-induction and hysteresis in, 99.
Telegraphy, multiple, 171; diplex, 171; contraplex, 171; quadruplex, 184.
Temperature, effect of, on action of voltaic cell, 20, 78; on resistance of substances, 58; effect of in increasing resistance of metals employed as telegraphic conductors, 78.
Temporary and permanent magnets, 24.
Tensile strengths of galvanized iron, hard and soft copper wires for telegraph lines, table of, 112.
Terminal station, arrangement of apparatus at, 158.
Terminal switchboard, 158.
Testing telegraph lines, 190; for disconnection at way-station, 134; by quantitative measurement, 195; character of measurements in, 195.
Tests of telegraph lines, object of, 190; of conductivity and insulation, recording of, 214.
Test-sheets, Western Union Company's forms for, 212, 213, 214.
Thermo-electricity, 3, 34.
Thompson, Silvanus P., on magnetic reactions, 31; on Wheatstone's bridge, 198.
Thompson's *Elementary Lessons in Electricity and Magnetism*, reference to, 31, 198; *Dynamo-Electric Machinery*, reference to, 86.
Thomson, Sir William, on importance of quantitative measurement in physical science, 36; *Popular Lectures and Addresses*, extract from, 36.
Thomson's theory of electricity, 2; cell, 17; rule for computing the windings of electro-magnets, 94.
Time, necessarily involved in formation of dot, 217; telegraphic code a multiple of, 216.
Transmitter, the, an element of the electric telegraph, 2.
Transmitter, pole changing, 181; single current, 173; single and double, how shown in diagram, 104.
Transverse magnetization, 26.
Trowbridge's *New Physics*, references to, 40, 85.
Twist-joint for iron wires, 113.
Tyndall, observations on conservation of energy, 38.
Tyndall's *Heat as a Mode of Motion*, reference to, 38.
Typical voltaic cell, description of, 3.

U OR HORSESHOE MAGNET, 26.
Unit of magnetism, 83; of magnetic induction, Kapp's, 83; English, 83.
Units, fundamental and derived, 37; electrical, synoptical table of, 73; practical, derived from natural constants, 59.
Universal switchboard, 156; manipulation of, 157.
U. S. Coast and Geodetic Survey, reference to reports of, 40.

VAIL, ALFRED, biographical notice of, 216; originator of the telegraphic register in its present form, 216; of the alphabetical code, 216.
Varley's insulator, tests of efficiency of, 120; loop test for position of escape on line, 203.
Varley's *Report on Western Union Lines*, references to, 131, 135.
V-gauge for wire, 95.
Victor key, 140.
Volt, the unit of electromotive force, definition of, 61; value of, 61.
Volta, Alessandro, biographical notice of, 61.
Volta induction, 74.
Voltaic cell, phenomena of, 5; directions for charging, 7.
Voltaic effect, chemistry of, 6.
Voltaic effect, the, 3.
Voltaic solutions, Becker's table of resistances of, 63; Farmer's ditto, 62.
Voltameter, the, 44.
Voltmeter, use of in telegraphic testing, 210; Weston's portable, 210; advantages of for testing, 211; Weston's combined ammeter and, for telegraphic testing, 210, 211.

Index.

WASHBURN & MORN M'F'G CO.'S IRON WIRE TABLE, 112.
Watt, James, biographical notice of, 73.
Watt, the unit of power or rate of work, value of, 73; rule for determination of, 73.
Way-station, arrangement of apparatus at, 151; connections of, 152; manipulation of switchboard in, 153; testing in, 154.
Weather-cross, term applied to escape between different wires, 132.
Webb, F. C., on electrostatic phenomena, 175.
Webb's *Electrical Accumulation and Conduction*, extract from, 175.
Weber, Eduard. 83.
Wedge or plug cut-out, the, 155.
Weights of iron and copper wires for telegraph lines, table of, 112.
Western Electric key, 140.
Western Union glass insulators, 116.
Western Union Telegraph Company's forms for test sheets, 212, 213, 214.
Western Union telegraph station, New York, arrangement of dynamos in, 170.
Weston's voltmeters and ammeters for telegraphic testing, 210, 211.
Wheatstone bridge, the, 195.
Wilkinson, Kennelly and, 85.
Winding magnet helices, machine for, 93.
Windings of helix, effect of position of, 92.
Windings of magnet coils, Sir Wm. Thomson's rule for computing, 94.

Wire-gauge, American standard, 95; British, 95.
Wire, instruments for gauging, 95, 96; in helix, relation of length and thickness of to number of turns, 92.
Wires, copper, for line construction, 114; for magnets, dimensions and resistances of, 93; Prescott's table of, 94.
Wires. crossing of, represented in diagram, 104.
Wires, iron, 111; table of sizes, weights, and resistances of, 112; joints in, 113; diagram of actual sizes of, 113.
Wood's repeater, 164.
Word-space, an element of telegraphic code, 216.
Work, electric, in circuit, 64.
Working efficiency of telegraph line, what it depends on, 127.

YOKE, of electro-magnet, 91.
Youmans' *Correlation and Conservation of Forces*, reference to, 38.

ZINC OF GRAVITY CELL, 4; process of cleaning, 16; amalgamation of, 21.
Zinc, oxide of, formed by action of voltaic cell, 6.
Zinc, electro-chemical equivalent of, 74, 75; chemical analysis of, 9.
Zinc solution of gravity cell, neutralization of, 16.
Zinc plate of cell the positive element, 12.
Zinc, d'Infreville's wasteless, 93.

LIST OF WORKS
ON
ELECTRICAL SCIENCE.
PUBLISHED AND FOR SALE BY
D. VAN NOSTRAND COMPANY,
23 Murray and 27 Warren Streets, New York.

ABBOTT, A. V. The Electrical Transmission of Energy. A Manual for the Design of Electrical Circuits. Illustrations and 9 folding plates. 8vo, cloth. $4.50.

ARNOLD, E. Armature Windings of Direct Current Dynamos. Extension and application of a general winding rule. Translated from the original German by Francis B. DeGress, M.E. (*In press.*)

ATKINSON, PHILIP. Elements of Static Electricity, with full description of the Holtz and Topler Machines, and their mode of operating. Second Edition. Illustrated. 12mo, cloth. $1.50.

The Elements of Dynamic Electricity and Magnetism. Third Edition. Illustrated. 12mo, cloth. $2.00.

Elements of Electric Lighting, including Electric Generation, Measurement, Storage, and Distribution. Eighth Edition. Fully revised and new matter added. Illustrated. 8vo, cloth. $1.50.

The Electric Transformation of Power and its Application by the Electric Motor, including Electric Railway Construction. Illustrated. 12mo, cloth. $2.00.

BADT, F. B. New Dynamo Tender's Handbook. 70 Illustrations. 16mo, cloth. $1.00.

Electric Transmission Handbook. Illustrations and Tables. 16mo, cloth, $1.00.

Incandescent Wiring Handbook. Fourth Edition. Illustrations and Tables. 12mo, cloth. $1.00.

Bell Hanger's Handbook. Third Edition. Illustrated. 12mo, cloth. $1.00.

BIGGS, C. H. W. First Principles of Electrical Engineering. Being an attempt to provide an Elementary Book for those who are intending to enter the profession of Electrical Engineering. Second Edition. Illustrated. 12mo, cloth. $1.00.

BLAKESLEY, T. H. Papers on Alternating Currents of Electricity. For the use of Students and Engineers. Third Edition, enlarged. 12mo, cloth. $1.50.

BOTTONE, S. R. Electrical Instrument-Making for Amateurs. A Practical Handbook. Sixth Edition. Enlarged by a chapter on "The Telephone." With 48 Illustrations. 12mo, cloth. 50 cents.

Electric Bells, and All about Them. A Practical Book for Practical Men. With over 100 Illustrations. Fifth Edition. 12mo, cloth. 50 cents.

The Dynamo: How Made and How Used. A Book for Amateurs. Eighth Edition. 100 Illustrations. 12mo, cloth. $1.00.

Electro-Motors: How Made and How Used. A Handbook for Amateurs and Practical Men. Illustrated. 12mo, cloth. 50 cents.

CLARK, D. K. Tramways: Their Construction and Working. Embracing a Comprehensive History of the System, with Accounts of the Various Modes of Traction, a Description of the Varieties of Rolling Stock, and Ample Details of Cost and Working Expenses; with Special Reference to the Tramways of the United Kingdom. Second Edition. Revised and rewritten. With over 400 Illustrations. Contains a special section on Electric Traction. Thick 8vo, cloth. $9.00.

CROCKER, F. B., and **WHEELER, S. S.** The Practical Management of Dynamos and Motors. Third Edition. Illustrated. 12mo, cloth. $1.00.

CROCKER, F. B. Electric Lighting. A Practical Exposition of the Art for the Use of Electricians, Students, and Others interested in the Installation or Operation of Electric Lighting Plants. Volume I.: The Generating Plant. 8vo, cloth. $3.00.

CUMMING, LINNÆUS, M.A. Electricity Treated Experimentally. For the Use of Schools and Students. Third Edition. 12mo, cloth. $1.50.

DESMOND, CHAS. Electricity for Engineers. Part I.: Constant Current. Part II.: Alternate Current. Revised Edition. Illustrated. 12mo, cloth. $2.50.

DU MONCEL, Count TH. Electro-Magnets: The Determination of the Elements of their Construction. 16mo, cloth. (No. 64 Van Nostrand's Science Series.) 50 cents.

DYNAMIC ELECTRICITY. Its Modern Use and Measurement, chiefly in its application to Electric Lighting and Telegraphy, including: 1. Some Points in Electric Lighting, by Dr. John Hopkinson. 2. On the Treatment of Electricity for Commercial Purposes, by J. N. Schoolbred. 3. Electric-Light Arithmetic, by R. E. Day, M.E. 18mo, boards. (No. 71 Van Nostrand's Science Series.) 50 cents.

EMMETT, WM. L. Alternating Current Wiring and Distribution. 16mo, cloth. Illustrated. $1.00.

EWING, J. A. Magnetic Induction in Iron and Other Metals. Second Issue. Illustrated. 8vo, cloth. $4.00.

FISKE, Lieut. BRADLEY A., U.S.N. Electricity in Theory and Practice; or, The Elements of Electrical Engineering. Eighth Edition. 8vo, cloth. $2.50.

FLEMING, Prof. J. A. The Alternate-Current Transformer in Theory and Practice. Vol. I.: The Induction of Electric Currents. 500 pp. Fifth Issue. Illustrated. 8vo, cloth. $3.00. Vol. II.: The Utilization of Induced Currents. Third Issue. 594 pp. Illustrated. 8vo, cloth. $5.00.

Electric Lamps and Electric Lighting. 8vo, cloth. $3.00.

FOSTER, HORATIO A. Electrical Engineer's Pocket-Book. (*In press.*)

GORDON, J. E. H. School Electricity. 12mo, cloth. $2.00.

GORE, Dr. GEORGE. The Art of Electrolytic Separation of Metals (Theoretical and Practical). Illustrated. 8vo, cloth. $3.50.

GUILLEMIN, AMÉDÉE. Electricity and Magnetism. Translated, revised, and edited by Prof. Silvanus P. Thompson. 600 Illustrations and several Plates. Large 8vo, cloth. $8.00.

GUY, ARTHUR F. Electric Light and Power, giving the result of practical experience in Central-Station Work. 8vo, cloth. Illustrated. $2.50.

HASKINS, C. H. The Galvanometer and its Uses. A Manual for Electricians and Students. Fourth Edition, revised. 12mo, morocco. $1.50.

Transformers: Their Theory, Construction and Application Simplified. Illustrated. 12mo, cloth. $1.25.

HAWKINS, C. C., and **WALLIS, F.** The Dynamo: Its Theory, Design, and Manufacture. 190 Illustrations. 8vo, cloth. $3.00.

HOBBES, W. R. P. The Arithmetic of Electrical Measurements. With numerous examples, fully worked. New Edition. 12mo, cloth. 50 cents.

HOSPITALIER, E. Polyphased Alternating Currents. Illustrated. 8vo, cloth, $1.40.

HOUSTON, Prof. E. J. A Dictionary of Electrical Words, Terms, and Phrases. Third Edition. Rewritten and greatly enlarged. Large 8vo, 570 illustrations, cloth. $5.00.

INCANDESCENT ELECTRIC LIGHTING. A Practical Description of the Edison System, by H. Latimer. To which is added: The Design and Operation of Incandescent Stations, by C. J. Field; A Description of the Edison Electrolyte Meter, by A. E. Kennelly; and a Paper on the Maximum Efficiency of Incandescent Lamps, by T. W. Howells. Illustrated. 16mo, cloth. (No. 57 Van Nostrand's Science Series.) 50 cents.

INDUCTION COILS: How Made and How Used. Fifth Edition. 16mo, cloth. (No. 53 Van Nostrand's Science Series.) 50 cents.

KAPP, GISBERT, C.E. Electric Transmission of Energy and its Transformation, Subdivision, and Distribution. A Practical Handbook. Fourth Edition, thoroughly revised. 12mo, cloth. $3.50.

Alternate-Current Machinery. 190 pp. Illustrated. (No. 96 Van Nostrand's Science Series.) 50 cents.

Dynamos, Alternators, and Transformers. Illustrated. 8vo, cloth. $4.00.

KEMPE, H. R. The Electrical Engineer's Pocket-Book: Modern Rules, Formulæ, Tables, and Data. Second Edition, with additions. 32mo, leather. $1.75.

A Handbook of Electrical Testing. Fifth Edition. 200 Illustrations. 8vo, cloth. $7.25.

KENNELLY, A. E. Theoretical Elements of Electro-Dynamic Machinery. Vol. I. Illustrated. 8vo, cloth. $1.50.

KILGOUR, M. H., and **SWAN, H.,** and **BIGGS, C. H. W.** Electrical Distribution: Its Theory and Practice. Illustrated. 8vo, cloth. $4.00.

LOCKWOOD, T. D. Electricity, Magnetism, and Electro-Telegraphy. A Practical Guide and Handbook of General Information for Electrical Students, Operators, and Inspectors. Fourth Edition. Illustrated. 8vo, cloth. $2.50.

LORING, A. E. A Handbook of the Electro-Magnetic Telegraph. 16mo, cloth. (No. 39 Van Nostrand's Science Series.) 50 cents.

MARTIN, T. C., and WETZLER, J. The Electro-Motor and its Applications. Fourth Edition. With an Appendix on the Development of the Electric Motor since 1888, by Dr. L. Bell. 300 Illustrations. 4to, cloth. $3.00.

MAVER, WM., Jr. American Telegraphy: Systems, Apparatus, Operations. 450 Illustrations. 8vo, cloth. 575 Pages. $3.50.

MORROW, J. T., and REID, T. Arithmetic of Magnetism and Electricity. 12mo, cloth. $1.00.

MUNRO, JOHN, C.E., and JAMIESON, ANDREW, C.E. A Pocket-Book of Electrical Rules and Tables. For the use of Electricians and Engineers. Eleventh Edition. Revised and enlarged. With numerous diagrams. Pocket size, leather. $2.50.

NIPHER, FRANCIS E., A.M. Theory of Magnetic Measurements. With an Appendix on the Method of Least Squares. 12mo, cloth. $1.00.

NOAD, H. M. The Student's Text-Book of Electricity. A New Edition. Carefully revised by W. H. Preece. 12mo, cloth. Illustrated. $4.00.

NOLL, AUGUSTUS. How to Wire Buildings. A Manual of the Art of Interior Wiring. Fourth Edition. 8vo, cloth. Illustrated. $1.50.

OHM, Dr. G. S. The Galvanic Circuit Investigated Mathematically. Berlin, 1827. Translated by William Francis. With Preface and Notes by the Editor, Thos. D. Lockwood. 12mo, cloth. (No. 102 Van Nostrand's Science Series.) 50 cents.

PALAZ, A. Treatise on Industrial Photometry. Specially applied to Electric Lighting. Translated from the French by G. W. Patterson, Jr., Assistant Professor of Physics in the University of Michigan, and M. R. Patterson, B.A. Second Edition. Fully Illustrated. 8vo, cloth. $4.00.

PARSHALL, H. F., and HOBART, H. M. Armature Windings of Electric Machines. With 140 full-page plates, 65 tables and descriptive letter-press. 4to, cloth. $7.50.

PERRY, NELSON W. Electric Railway Motors. Their Construction, Operation, and Maintenance. An Elementary Practical Handbook for those engaged in the management and operation of Electric Railway Apparatus, with Rules and Instructions for Motormen. 12mo, cloth. $1.00.

PLANTÉ, GASTON. The Storage of Electrical Energy, and Researches in the Effects created by Currents combining Quantity with High Tension. Translated from the French by Paul B. Elwell. 89 Illustrations. 8vo. $4.00.

POOLE, J. The Practical Telephone Handbook, and Guide to the Telephonic Exchange. Second Edition. Revised and enlarged. Illustrated. 8vo, cloth. $1.50.

POPE, F. L. Modern Practice of the Electric Telegraph. A Handbook for Electricians and Operators. An entirely new work, revised and enlarged, and brought up to date throughout. Illustrations. 8vo, cloth. $1.50.

PREECE, W. H., and STUBBS, A. J. Manual of Telephony. Illustrated. 12mo, cloth. $4.50.

RECKENZAUN, A. Electric Traction. Illustrated. 8vo, cloth. $4.00.

RUSSELL, STUART A. Electric-Light Cables and the Distribution of Electricity. 107 Illustrations. 8vo, cloth. $2.25.

SALOMONS, Sir DAVID, M.A. Electric-Light Installations. A Practical Handbook. Seventh Edition, revised and enlarged. Vol. I.: Management of Accumulators. Illustrated. 12mo, cloth. $1.50. Vol. II.: Apparatus. Illustrated. 12mo, cloth. $2.25. Vol. III.: Application. Illustrated. 12mo, cloth. $1.50.

SCHELLEN, Dr. H. Magneto-Electric and Dynamo-Electric Machines. Their Construction and Practical Application to Electric Lighting and the Transmission of Power. Translated from the third German edition by N. S. Keith and Percy Neymann, Ph.D. With very large Additions and Notes relating to American Machines, by N. S. Keith. Vol. I. with 353 Illustrations. Third Edition. $5.00.

SLOANE, Prof. T. O'CONOR. Standard Electrical Dictionary. 300 Illustrations. 8vo, cloth. $2.50.

SNELL, ALBION T. Electric Motive Power. The Transmission and Distribution of Electric Power by Continuous and Alternate Currents. With a Section on the Applications of Electricity to Mining Work. Illustrated. 8vo, cloth. $4.00.

SWINBURNE, JAS., and WORDINGHAM, C. H. The Measurement of Electric Currents. Electrical Measuring Instruments. Meters for Electrical Energy. Edited, with Preface, by T. Commerford Martin. Folding Plate and numerous Illustrations. 16mo, cloth. 50 cents.

THOM, C., and JONES, W. H. Telegraphic Connections, embracing recent methods in Quadruplex Telegraphy. Twenty colored plates. 8vo, cloth. $1.50.

THOMPSON, EDWARD P. How to Make Inventions; or, Inventing as a Science and an Art. An Inventor's Guide. Second Edition. Revised and Enlarged. Illustrated. 8vo, paper. $1.00.

THOMPSON, Prof. S. P. Dynamo-Electric Machinery. With an Introduction and Notes by Frank L. Pope and H. R. Butler. Fully Illustrated. (No. 66 Van Nostrand's Science Series.) 50 cents.

Recent Progress in Dynamo-Electric Machines. Being a Supplement to "Dynamo-Electric Machinery." Illustrated. 12mo, cloth. (No. 75 Van Nostrand's Science Series.) 50 cents.

The Electro-Magnet and Electro-Magnetic Mechanism. Second Edition, revised. 213 Illustrations. 8vo, cloth. $6.00.

TREVERT, E. Practical Directions for Armature and Field-Magnet Winding. Illustrated. 12mo, cloth. $1.50.

How to Build Dynamo-Electric Machinery. Embracing the Theory, Designing, and Construction of Dynamos and Motors. With Appendices on Field-Magnet and Armature Winding, Management of Dynamos and Motors, and useful Tables of Wire Gauges. Illustrated. 8vo, cloth. $2.50.

TUMLIRZ, Dr. Potential, and its Application to the Explanation of Electrical Phenomena. Translated by D. Robertson, M.D. 12mo, cloth. $1.25.

TUNZELMANN, G. W. de. Electricity in Modern Life. Illustrated. 12mo, cloth. $1.25.

URQUHART, J. W. Dynamo Construction. A Practical Handbook for the Use of Engineer Constructors and Electricians in Charge. Illustrated. 12mo, cloth. $3.00.

Electric Ship-Lighting. A Hand-book on the Practical Fitting and Running of Ships' Electrical Plant, for the Use of Ship Owners and Builders, Marine Electricians and Sea-going Engineers in Charge. 88 Illustrations. 12mo, cloth. $3.00.

Electric Light Fitting. A Hand-book for Working Electrical Engineers, Embodying Practical Notes on Installation Management. Second Edition, with additional chapters. With numerous Illustrations. 12mo, cloth. $2.00.

WALKER, FREDERICK. Practical Dynamo-Building for Amateurs. How to Wind for any Output. Illustrated. 16mo, cloth. (No. 98 Van Nostrand's Science Series.) 50 cents.

WALMSLEY, R. M. The Electric Current. How Produced and How Used. With 379 Illustrations. 12mo, cloth. $3.00.

WEBB, H. L. A Practical Guide to the Testing of Insulated Wires and Cables. Illustrated. 12mo, cloth. $1.00.

WORMELL, R. Electricity in the Service of Man. A Popular and Practical Treatise on the Application of Electricity in Modern Life. From the German, and edited, with copious additions, by R. Wormell, and an Introduction by Prof. J. Perry. With nearly 850 Illustrations. Royal 8vo, cloth. $5.00.

WEYMOUTH, F. MARTEN. Drum Armatures and Commutators. (Theory and Practice.) A complete treatise on the theory and construction of drum-winding, and of commutators for closed-coil armatures, together with a full *résumé* of some of the principal points involved in their design; and an exposition of armature reactions and sparking. Illustrated. 8vo, cloth. $3.00.

www.ingramcontent.com/pod-product-compliance
Lightning Source LLC
Chambersburg PA
CBHW021407230426
43666CB00006B/669